This short report has been written by:—

- Patricia Murphy
- Beta Schofield

D1796460

Much of the work on which it is based is the product of planning and development in which all members of the research teams, listed below, and the Steering Group on Science (listed on page 42) have had a part. In particular the report draws on the full reports of the first and second surveys of pupils aged 13 which were written jointly by Beta Schofield, Paul Black, Richard Gott, Jenny Head, Sandra Johnson and Patricia Murphy.

## The present science teams

### Chelsea

| | |
|---|---|
| Director | Paul Black |
| Research and development (ages 11 and 13) | Wynne Harlen |
| | Patricia Murphy |
| | Tony Orgee |
| | David Palacio |
| | Terry Russell |
| | Beta Schofield |
| Secretarial staff | Peggy Walker |
| | Margaret Hunt |
| Technician | Barrie Downer |

### Leeds

| | |
|---|---|
| Director | Fred Archenhold |
| Technical director | Roger Hartley |
| Statistics | Sandra Johnson |
| | Jenny Head |
| | John Bell |
| | Nasrin Khaligh |
| Research and development (age 15) | Richard Gott |
| | Angela Davey |
| | Reed Gamble |
| | Geoff Welford |
| Secretarial staff | Glynis Wilkinson |
| | Helen Johnson |
| | Elizabeth Lodge |

### Past team members

| | |
|---|---|
| David Layton | (Director 1977-82) |
| Dennis Child | (Director 1982-83) |
| Brenda Denvir | (Jan.-Aug. 1981) |
| Rosalind Driver | (1977-82) |
| Brian Maher | (1980-83) |
| Cynthia Millband | (1980-81) |
| Ardrie VanderWaal | (1979-80) |
| Christopher Worsley | (1978-82) |
| Fiona Wylie | (1980-82) |

N.F.E.R.
INFORMATION
SERVICE LIBRARY

DATE: 30.4.84

CLASS: JFO MUR

ACC. NO: 26649

A. CARD MURPHY P

S. CARD

1

# CONTENTS

# INTRODUCTION

## What this report is about

This is an account of some of the results of the national surveys which were carried out in 1980 and 1981 to find out about 13 year old pupils' performance in science. It includes an outline of the assessment framework, some of the questions which were written to match it, a description of how well, and how differently, pupils responded to them and suggests how the information might be used.

The assessment is concerned with what 13 year olds as a whole can do, not what particular pupils can do, and in this way is quite different from ordinary examinations. Each year a number of randomly sampled schools of all types are invited to take part on a voluntary basis. In 1980, 448 English schools were involved; in 1981 the survey was extended to cover Wales and Northern Ireland and the number of schools rose to 556. In general about 27 pupils are randomly selected (by date of birth) from each school. These pupils make up rather less than 2% of the 13 year old population. Individual pupils, schools or LEAs cannot be identified with particular results; and the names of pupils taking part are not known outside their own school. Between 20 and 30 different packages of questions are used in each survey, and no pupil takes more than two of them; thus it is not until the results are brought together that a meaningful picture of performance in science emerges.

## The assessment framework

Before considering the results of the survey, readers need to be aware of the assessment framework[1], for it is to this which questions used in the survey owe their special characteristics. The framework, in turn, reflects a particular view of science; this has been chosen from several possible alternatives because it is appropriate for the pupils of the relevant ages and of all abilities; and because the framework to which it gives rise allows survey results to be expressed in ways which are informative and useful to a very wide audience.

Science is seen as an experimental subject concerned with problem-solving _ a complex activity involving the operation of a set of processes as well as the understanding of a body of scientific concepts and knowledge. Among the categories assessed, 'Performance of Investigations' (Category 6) is the most complete expression of this view. However, it is also useful to examine and report on various aspects of this scientific activity separately, and these aspects form the subjects of the other five categories.

Categories of Science Performance

| | |
|---|---|
| 1. Use of graphical and symbolic representation (w) | Reading information from graphs, tables & charts. Representing information as graphs, tables & charts. |
| 2. Use of apparatus and measuring instruments (p) | Using measuring instruments. Estimating physical quantities. Following instructions for practical work. |
| 3. Observation (p) | Making and interpreting observations. |
| 4. Interpretation and application (w) | I   Interpreting presented information. II  Applying:  biology concepts<br>                  physics concepts<br>                  chemistry concepts |
| 5. Planning of investigations (w) | Planning parts of investigations. Planning entire investigations. |
| 6. Performance of investigations (p) | Performing entire investigations. |

(w) = written tests    (p) = practical tests

Any questions used in the assessment fits into only one of the sub-categories. Thus performance can be described for each in turn, either by giving an overall mean score for all questions representing it in the survey, or by detailed accounts of responses to individual questions, or both. A few other aspects of science were also assessed in 1980 and 1981, and are discussed under the appropriate category heading.

The results for any particular question may be described as follows:

*The percentages of pupils responding in certain defined ways, whether correct or not, may be stated.

*The percentages of pupils awarded different scores, from zero to a maximum for the question, may be presented, usually as a histogram.

*The mean score for the question may be given; this is the average of the scores of all the pupils. (If a question carries only one mark, its mean score is the same as the percentage of pupils giving the correct response.)

Although it is easy enough to write down the names of sub-categories concerned with different activities of science, it has to be remembered that it is not possible to completely isolate those activities in practice. There will be a degree of interpretation required, a touch of observation, an element of following instructions, whatever the question to be answered.

4

# PERFORMANCE OF INVESTIGATIONS
## (Category 6)

The example used to illustrate this category is called 'Survival' and is shown on page 6. Even though they are given the same question pupils' perceptions of the problem and the strategies they adopt for its solution vary considerably; consequently a wide range of equipment is provided in case it is needed.

Each tester sees one pupil at a time. The equipment for the investigation is presented to each pupil in exactly the same way. The tester gives a standard introduction[2] explaining the nature of the test and what the pupil has to do, and describes the equipment including how each piece of apparatus works. The pupils are first asked to read out the question in the box. If it is necessary it is read out for them. The pupils are told that it is up to them to decide what apparatus to use and that they do not have to use it all. They are then asked to read the question through again and to tell the tester, in their own words, what they think they have to find out. No indication is given to the pupil, over and above that already described, of what the problem is or how the apparatus can be used to solve it. About half an hour per question is allowed, during which the pupils plan, carry out, evaluate and possibly do a trial run or repeat their experiments.

The **aim of this assessment** is to observe the route the individual pupil takes to the solution and to assess the quality of it. The emphasis is on what the pupil decides to do: there is no *one* correct route. The pupils are given a piece of paper and asked to write down their results only, not what they do. Each pupil's actions are recorded on a checklist designed in such a way that the experiment can subsequently be reconstructed in considerable detail.

**Analysis and results for 'Survival'**
Pupils use many routes to solve this problem. Some of them, although different, are equivalent in terms of deriving an adequate solution. As a consequence the analysis has attempted to describe the variety of response and to categorize the adequacy of it rather than to give a score which would not be very informative. The checklist method of recording the pupil's **actions** is best explained by referring to the example shown on pages 5 and 6. Detailed instructions for coding the checklists are issued during the training session which is attended by experienced teachers before they administer the test[2].

In this question the *independent variable* to be tested is the type of fabric. The *dependent variable* to be judged is the thermal conductivity of the fabrics.

The problem is set outside the laboratory situation but the pupil is given clues about how to simulate the situation using laboratory apparatus. The apparatus presented to the pupils and its layout are shown in the pictures.

> Imagine you are stranded on a mountainside in cold, dry, windy weather. You can choose a jacket made from one of the fabrics in front of you.
>
> This is what you have to find out:-
>
> > Which fabric would keep you warmer?
>
> You can use any of the things in front of you. Choose whatever you need to answer the question.
>
> You can:
>
> - use a tin instead of a person
>
> - put warm water inside to make it more life-like
>
> - make it a 'jacket' from the material
>
> - use a hair dryer to make an imitation wind
>   (without the heater switched on, of course!)
>
> Make a clear record of your results so that someone else can understand what you have found out.

*Apparatus:* The pupils are presented with the range of apparatus indicated in the picture. They have several alternatives for fastening the fabric: pins, rubber bands and sellotape. They are given 5 cans — 4 metal (2 of which are identical in size) and 1 plastic can of this same size. They also have a hairdryer and stand to use if they wish to accelerate the cooling process or simulate the wind.

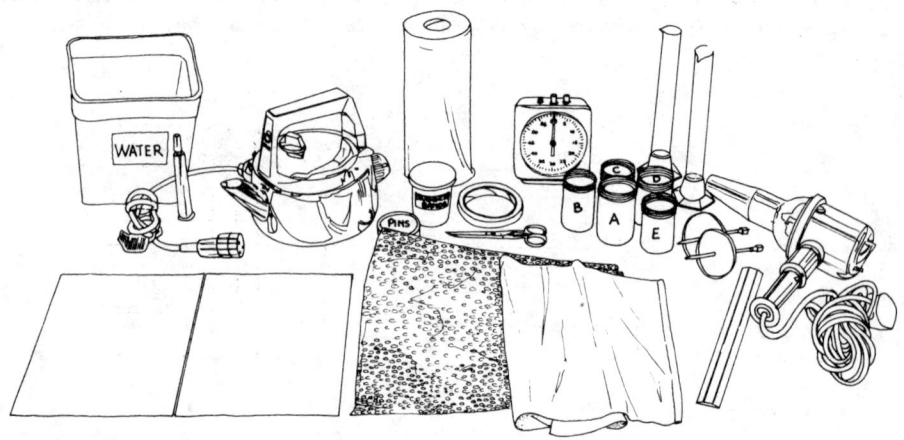

# The Checklist for 'Survival'

The activities listed allow for different approaches and different ways of setting up and performing the investigation. In the shaded rows, the testers enter the numerical values of quantities which they themselves observe while the investigation is in progress.

The ten columns on the checklist allow for the recording of ten **trials**. A trial represents any experiment carried out by the pupil on one of the materials.

TRIALS

| | 1 | 2 | 3 | 4 | 5 | 6 | 7 | 8 | 9 | 10 | |
|---|---|---|---|---|---|---|---|---|---|---|---|
| Material used - blanket | | | | | | | | | | | |
| - plastic | | | | | | | | | | | |
| Material around - hand | | | | | | | | | | | |
| - thermometer | | | | | | | | | | | |
| - can A | | | | | | | | | | | |
| B | | | | | | | | | | | |
| C | | | | | | | | | | | |
| D | | | | | | | | | | | |
| E | | | | | | | | | | | |
| Used material for $<$ 1 layer | | | | | | | | | | | |
| $>$ 1 layer | | | | | | | | | | | |
| - to cover base | | | | | | | | | | | |
| Material fixed in place | | | | | | | | | | | |
| Material held in place | | | | | | | | | | | |
| Material used alone | | | | | | | | | | | |
| Used hot water ($>$ 60°C) in can | | | | | | | | | | | |
| Used warm water (35°-60°) in can | | | | | | | | | | | |
| Used cold water ($<$ 35°C) in can | | | | | | | | | | | |
| Water measured by cylinder | | | | | | | | | | | |
| Water measured by eye | | | | | | | | | | | |
| *Actual volume (cm$^3$) | | | | | | | | | | | |
| Read Initial temp. of water before cooling period | | | | | | | | | | | |
| Actual temp. (° C) | | | | | | | | | | | |
| Started clock | | | | | | | | | | | |
| - within $\pm$5s of reading temp. | | | | | | | | | | | |
| Records temp. at regular intervals | | | | | | | | | | | |
| Number of records: 2 | | | | | | | | | | | |
| $>$ 2 | | | | | | | | | | | |
| Read final temp. of water | | | | | | | | | | | |
| Read final temp. after set time | | | | | | | | | | | |
| Read final temp. after set temp. drop | | | | | | | | | | | |
| Time interval $<$ 2 minutes | | | | | | | | | | | |
| $>$ 5 minutes | | | | | | | | | | | |
| Actual time (s) | | | | | | | | | | | |
| Actual final temp. (°C) | | | | | | | | | | | |
| Used hairdryer - alternating between cans | | | | | | | | | | | |
| - directed between cans | | | | | | | | | | | |
| - on one material only | | | | | | | | | | | |
| Range used - touching material | | | | | | | | | | | |
| - $<$ 5 cm | | | | | | | | | | | |
| - $>$ 5 cm | | | | | | | | | | | |
| Trial/s used in record to conclude | | | | | | | | | | | |
| Approach used in each trial with record | | | | | | | | | | | |

*After the investigation is over,* the pupils' performance, written record and spoken answers to the four questions below are considered by the testers in conjunction with the numerical values they themselves noted. The pupils' spoken responses to the question are recorded elsewhere by the testers. These constitute the **spoken record**.

The pupils' actions, already recorded on the checklist, together with the testers' own recorded values, form the **evidence**.

**Control** of a variable is said to have occurred if the testers' recorded values of the quantity are the same with +/-10%.

SURVIVAL

Administrator ☐   Sex ☐          Pupil No. ☐☐☐☐
Date of testing ☐☐☐☐            School No. ☐☐☐☐
Time am/pm ☐                      D of B ☐☐☐☐
No. of session ☐                 Sex B/G ☐    Order ☐
                                 Curriculum ☐☐

QUESTIONS

1) What have you found out?
2) How did you tell which fabric would keep you warmer?
3) What did you do to make sure it was a fair test?
4) If you had to do this again, is there anything you would do differently?

RECORD

| | ALL TRIALS | SOME TRIALS | CONCLUSION TRIALS |
|---|---|---|---|
| Spoken record consistent with evidence | ☐ | ☐ | ☐ |
| Written record consistent with evidence | ☐ | ☐ | ☐ |
| Spoken conclusion consistent with evidence | ☐ | ☐ | ☐ |
| Written conclusion consistent with evidence | ☐ | ☐ | ☐ |

No written conclusion ☐

| Written record specifies: | ALL | PART |
|---|---|---|
| type of fabric | ☐ | ☐ |
| quantities measured | ☐ | ☐ |
| correct units | ☐ | ☐ |

Answer given (blanket or plastic) ☐
( B        P )
Style:  tabulated ☐
        ordered ☐
        random ☐

CONTROL

| | CAN | FABRIC | INITIAL TEMP. | VOL. WATER | FAST-ENING | COOLING |
|---|---|---|---|---|---|---|
| Control for no trials | ☐ | ☐ | ☐ | ☐ | ☐ | ☐ |
| Control for enough trials to conclude from but fails to use them | ☐ | ☐ | ☐ | ☐ | ☐ | ☐ |
| Control for conclusion trials | ☐ | ☐ | ☐ | ☐ | ☐ | ☐ |
| Control for all trials | ☐ | ☐ | ☐ | ☐ | ☐ | ☐ |

COMMENTS

Method is uncodable (write on separate sheet) ☐
Pupil's performance badly affected by stress ☐
Thermometer used correctly (in the water) ☐

8

The analysis of the investigation reflects the stages in the problem-solving chain which is discussed in the Assessment Framework.[1] Each experiment can be seen as being made up of the following components.

- The adoption of an overall **approach** — this is an initial selection which may be modified by the pupil during the experiment.
- The **setting up** of the experiment — this includes establishing the initial conditions, e.g. the temperature of the water used; which can or cans were selected, etc.
- The actual **measurement** related to the dependent variable which allows judgements to be made of that variable — this includes details such as time intervals, numbers of measurements taken, etc.
- **The collection and recording** of data.
- The **evaluation** procedures — these reflect the degree and type of evaluation occurring, by what proportion of pupils.
- The **control of variables** — this includes both the degree and type of controls exercised.

## Approach

To determine the **approach** used by each pupil, the data were searched by computer for two trials — one trial on each material — which were identical and therefore could be used to effect a comparison. If there were more than two trials, the best two were selected.

'Survival' Approaches (Number of pupils = 618)

| | The % of pupils for which approach was used: |
|---|---|
| A. Fabric wrapped around hand — qualitative | 3 |
| B. Fabric around thermometer — qualitative | 0 |
| C. Fabric around thermometer — set temperature drop time interval measured | 0 |
| D. Fabric around thermometer — set time interval: temperature drop measured | 1 |
| E. Fabric around can and water — qualitative | 11 |
| F. Fabric around can and water — set temperature drop: time interval measured | 3 |
| G. Fabric around can and water — set time interval: temperature drop measured | 69 |
| H. Fabric around can and water — same final temperature: time interval measured | 1 |
| I. Final temperature measured — not timed | 9 |

*In order to tackle the question adequately a quantitive measurement of the rate of cooling, for both materials, had to be made. 80% of the pupils who attempted this investigation used a quantified approach. However 19% who used approach F, G or H failed to take an initial temperature reading.*

**Setting up:** Once the pupils have decided on their general approach and what to measure, they have to work out the practical details necessary to set up the experiment adequately. An experimental set-up which takes all the initial conditions into account is called level one and successive lower levels define the conditions with some factors missing. These levels are then applied to the checklist data to describe how well the pupils set up the experiment.

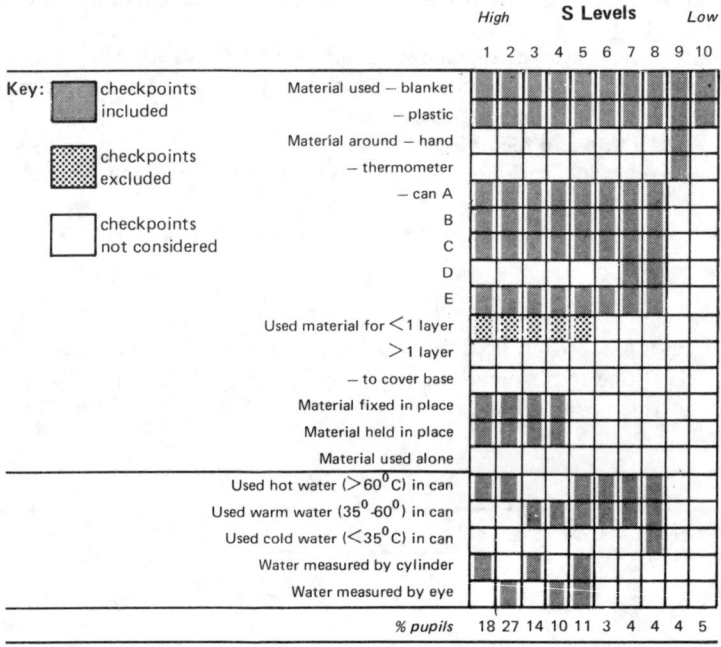

Overall about 70% of the pupils set up the experiment taking most of the variables — the can, the fabric and the fastening — into account. 18% of the pupils set up an experiment where all the initial conditions were taken into account, including a carefully measured amount of water. A further 27% did everything but judged the volume of water by eye rather than by cylinder. 24% of the pupils used warm water rather than hot water but established all the other conditions: whilst this is an understandable approach it did not allow for a drop in temperature of sufficient magnitude.

**Measurement:** The measurement techniques employed by the majority of pupils in the operationalisation of the dependent variable (i.e. the rate of cooling) suggest that they have little experience in this area of problem-solving. The pupils knew on the whole what to measure but procedures which affect the validity and accuracy of the measurement were generally unknown at this age. Only 7% of the pupils allowed a time interval of more than 5 minutes for cooling to occur; 60% of the pupils used a time interval of less than 2 minutes. Only 3% took the temperature at regular intervals and 5% made more than two records of the temperature as it fell.

| | High 1 | 2 | 3 | 4 | 5 | 6 | 7 | 8 | 9 | 10 Low | |
|---|---|---|---|---|---|---|---|---|---|---|---|
| *Actual volume (cm$^3$) | | | | | | | | | | | |
| Read initial temp. of water before cooling period | | | | | | | | | | | |
| Actual temp. ($^0$C) | | | | | | | | | | | N |
| Started clock | | | | | | | | | | | O |
| – within ±5s of reading temp. | | | | | | | | | | | |
| Records temp. at regular intervals | | | | | | | | | | | M E A |
| Number of records: 2 | | | | | | | | | | | |
| >2 | | | | | | | | | | | S |
| Read final temp. of water | | | | | | | | | | | U R |
| Read final temp. after set time | | | | | | | | | | | E |
| Read final temp. after set temp. drop | | | | | | | | | | | M |
| Time interval <2 minutes | | | | | | | | | | | E N |
| l >5 minutes | | | | | | | | | | | T |
| Actual time (s) | | | | | | | | | | | |
| Actual final temp. ($^0$C) | | | | | | | | | | | |
| % pupils | 1 | 2 | 2 | 17 | 12 | 9 | 7 | 32 | 17 | | |

**A comparison of performance in 'setting up' and 'measurement'**

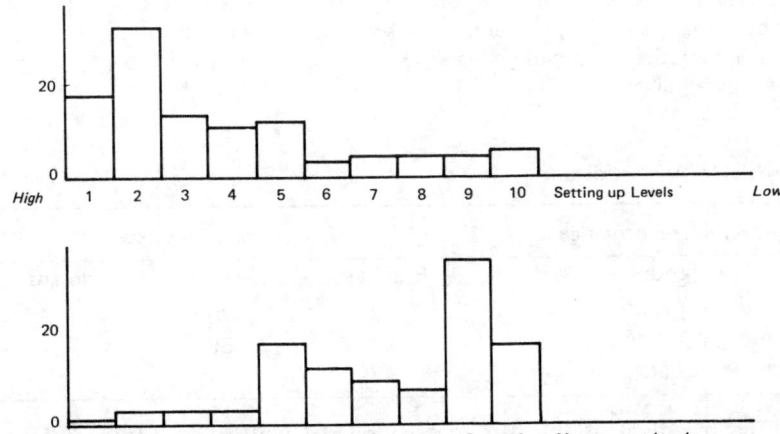

% pupils   High 1 2 3 4 5 6 7 8 9 10   Setting up Levels   Low

1 2 3 4 5 6 7 8 9 10   Measurement Levels

## Overall performance

Three overall levels of performance were established for each of the stages **approach**, **setting up** the experiment, and **measurement**.

'Survival': Overall Performance Groups

| Overall | | Permitted levels of: | | | |
|---|---|---|---|---|---|
| performance group | Approach | Setting up | Measurement | % pupils* |
| G1 | } F G H | 1 - 2 | 1 - 5 | 9 |
| G2 | | 1 - 7 | 1 - 7 | 34 |
| G3 | A - I | 1 - 10 | 1 - 10 | 49 |

*The remaining 8% did not test both fabrics.

In Group **G1** pupils must have tested both the blanket and plastic; used at least one layer of fabric fastened or held around a metal can containing hot water; read the initial temperature of the water and timed the cooling interval using a clock, allowing at least two minutes before the final temperature of the water was taken.

In Group **G2** were those pupils who tested both the blanket and the plastic around a metal can and either used hot or warm water and recorded both an initial and a final temperature. The cooling time did not have to be measured by the clock and its duration could be for less than two minutes.

In Group **G3** the minimum requirement was that the independent variables — blanket and plastic — were identified. 33% of the total number of pupils tested who fell into this group did attempt to make measurements related to the dependent variable. These pupils tested both the blanket and plastic around cans (either metal or plastic) containing water. The water used could be either hot, warm or cold. No initial temperature reading was taken but a final temperature was recorded.

## Collecting and recording data

The emphasis of the assessment was on the method of problem-solving, but it was also important to consider the solution to the problem. The relatiionship between the level of performance of the pupil and the answer given was investigated. The better insulator was, in fact, the plastic; the table shows the % of pupils in each performance group and the answers given.

'Survival': Answer given

| Overall performance | % of pupils | | |
|---|---|---|---|
| group | Blanket | Plastic | No answer |
| **G1** ( 9%) | 27 | 70 | 3 |
| **G2** (34%) | 37 | 56 | 7 |
| **G3** (49%) | 42 | 51 | 7 |

The recording of the data was considered with respect to its consistency with evidence, its content, and style. On all these aspects of record-keeping the pattern was the same — those pupils in group **G1** consistently performed best.

## Evaluation and modifications

The vast majority of the pupils who attempted the investigation and tested two fabrics (70%) used two trails only and drew their conclusions from them. The 20% of pupils who used four trials rejected two, and these trials indicated how the pupils modified their actions. The most common changes were those in which the pupils modified their experimental set-up. Very few changes related to improvements in measurement procedures or to a change in general approach. 18% of the changes, however, indicated that pupils within the experimental framework recognised a need to make measurements which previously had not been deemed appropriate or necessary.

## Control of variables

Control of variables is not considered when making a judgement of the overall level of performance of an individual pupil, for control of variables can only be judged by looking across trials and is assessed after completion of the investigation. The judgement made by the tester is based on the actions recorded for the pupil and on the numerical values observed by the tester (i.e. the actual amount of fabric used, the volume of water in the can, etc.). The variables which should be kept the same across trials in this investigation with the percentages of pupils who did control them are shown in the table below.

'Survival': Control of variables

| Variables to be controlled | % of pupils |
| --- | --- |
| Can (both size and material) | 67 |
| Fabric (size) | 82 |
| Initial temperature of the water | 38 |
| Volume of water | 51 |
| Fastening of the fabric | 85 |
| Conditions of cooling of the cans | 60 |

## Other investigations

Other questions quite different from 'Survival' were also used in the survey. Each one had to be appropriate for scientific experimentation and able to be attempted and assessed within 45 minutes.

Two aspects of investigations, the context and the type, were considered to affect pupil performance. In the first case, for example, if the pupils' problem involved assessing the behaviour of live animals rather than measuring physical quantities, certain actions become more or less appropriate. Secondly the type of investigation affects the demands made on the pupils in terms of the component skills from the other categories that are required. The six questions used in 1981 were selected because of their differences both in context and in their demands with respect to problem-solving.

## Difference between investigations

**The number of possible approaches** varied from question to question and was determined by the number and nature of the independent and dependent variables and also the variety of methods available for judging the dependent variable, which can be distinguished from that which they decide to measure in order to inform their judgements. The approaches considered for each investigation were worked out logically in the first instance and then modified in the light of small-scale trials to include all the different ways the pupils perceived the problem.

### 1981 Question summary

| Question | The variable to be varied systematically i.e. the independent variable | The variable to be judged i.e. the dependent variable | Number of approaches considered |
|---|---|---|---|
| SURVIVAL: Which fabric keeps you warmer? | Fabric (x2) | thermal conductivity | 12 |
| CARS: If all the cars are given the same chance, which one will travel furthest? | Cars (x3) | Distance travelled | 9 |
| WOODLICE: If woodlice are given the choice of the four places below, which do they choose to live in — *Damp* and *Dark* or *Dry* and *Dark* or *Damp* and *Light* or *Dry* and *Light?* | Dampness Darkness | Woodlice "happiness" | 12 |
| HOT WASH: Does this washing powder wash dirty cloth as clean in cold water as it does in hot water? | Cloth dirt Water temperature | Cleanness | 8 |
| PAPER TOWEL: Which kind of paper will hold the most water? | Paper (x3) | Water held | 12 |
| FLOORING: Which one of the floor coverings do you think is the most suitable for a kitchen floor? | Floor covering (x4) | Suitability | 7 |

By choosing different kinds of investigation the experiments can be arranged to spread thinly across all aspects of the problem solving chain or cluster around particular events in the chain.[1]

# Summary: PERFORMANCE OF INVESTIGATIONS

**Diversity:** Looking at the results across investigations it is clear that most 13 year olds are successful in setting up investigations to solve problems; they show enthusiasm and an unexpected diversity of approach.

**Context:** Pupils can and do operate on the same question quite differently when set in a **practical** context and when in a **written** context. They are much more likely to adopt 'scientific' solutions in the practical context: the opportunity for practically working through a problem results in a more realistic and successful path to the solution being adopted. The context of individual questions determines what the pupils see as appropriate solutions; an everyday setting or one which is familiar to the pupil may lead the pupil away from the 'scientist's' quantitative approach.

**Measurement:** Once a problem has been put into an experimental set-up obstacles to success arise subsequently during the carrying out of the investigation, for example when measurement procedures and techniques are required. Other obstacles appear to be specific to particular investigations. These may arise when understanding of science concepts has to be brought to bear e.g. insulators in 'Survival', or when the demands of a particular investigation are complex. In particular pupils may have difficulty in the identification of the independent variables when there are more than one, as in 'Woodlice'.

**Control of variables:** Which variables the pupils control is related to the 'obviousness' of the variable, e.g. fabric size in 'Survival' (82%), and to the degree of conceptual understanding needed to decide whether control is necessary, e.g. volume of water in 'Survival' (51%).

How many variables are controlled is linked to overall performance in that the better the pupils' experiments then the more likely they are to have taken extraneous variables into account; for example, 62% of pupils in **G1** compared to 56% in **G2** and 46% in **G3** controlled the volume of water in 'Survival'. Pupils are asked at the end of the experimental period what they did to make sure it was a fair test. Pupil responses show that they often control a variable and then do not consider it worth reporting. For example 51% controlled the volume of water but only 24% said that they had done so. This suggests that pupils are not, in fact identifying the variables that require control. Rather they are exercising control as part of the process of carefully carrrying out a scientific investigation where things should be the same for a fair test.

# PLANNING OF INVESTIGATIONS (Category 5)

To test "planning" the questions are described in writing, but are sometimes accompanied by a picture which may be purely illustrative or which may be used to convey additional information. In this category the pupils are in a situation which is remote from the practical context and yet unrestricted with respect to apparatus and procedures. Three types of activity were assessed in this category:
- **Planning entire investigations**
- **Proposing testable statements**
- **Planning parts of an investigation**

## Planning entire investigations

In this part of the assessment the pupils' responses are extended, occupying them for about 15 minutes. The analysis features are those considered in Category 6, so that 'planning' and 'doing' can be compared.

'Survival' : *'Planning' and 'doing' compared*

Tony and Brian are given two pieces of material, each about twice as big as this sheet of paper. One piece is <u>blanket</u> and the other is <u>plastic</u>.

They decide to find out <u>which material is better for keeping people warm</u>.

To do this they make a cover from each type of material and put them around cans of hot water.

Describe the experiment they should do to test which material is better. Be sure to write about:

- how to make the covers
- the equipment you would use
- the things you would measure
- the things you would do to make it a fair test
- how you would work out your results

*This example is a question to test planning based on the same content as the earlier question on 'Survival', which was a test of 'doing'.*

**Planning** — *Category 5*
*In the planning question above the pupils were told explicitly that the materials were to be placed around cans of hot water and the context was science. In Category 5 only the pupils' stated intent can be taken into account. The pupils have no results to assess and no opportunity to reconsider their experimental design.*

**Doing** — *Category 6*
*In the question to test performing the investigation, suggestions were made to enable pupils to simulate the real situation. All the pupils' actions are taken into account, even those that they themselves later reject or ignore. The pupils have to set up apparatus and make measurements and there is scope for modifying their experiment.*

**The two categories therefore represent very different activities in spite of the obvious similarities between the questions.**

## Analysis

The pupil's plan is considered to see whether the **independent variable** – the type of **fabric**, was identified, and what criterion was used to assess the **thermal conductivity** i.e. the **dependent variable.**

| 'Survival planned' approaches          n = 823 | % of pupils Planning | % of pupils Doing |
|---|---|---|
| - Approach | Planning | Doing |
| A. Fabric around hand/body – qualitative | 3 | 4 |
| B. Fabric around thermometer – qualitative | 2 | 0.5 |
| C. Fabric around can and water – qualitative | 25 | 11 |
| D. Fabric around can and water – time measured for set temperature drop | 3 | 3 |
| E. Fabric around can and water – temperature drop measured for a set time | 48 | 69 |
| F. Fabric around can and water – time measured until the same final temperature reached | 2 | 1 |
| G. Only one fabric considered | 2 | 2 |
| Category 5 only | | |
| J. Reformulation | 4 | |
| K. States answer without devising test | 5 | |
| L. Irrelevant/incomprehensible | 5 | |
| P. No approach described | 3 | |
| N. No response | 5 | |

**Planning** – *Category 5*
*53% of pupils used an acceptable approach.*
*21% failed to identify either the independent or the dependent variable.*

**Doing** – *Category 6*
*73% of pupils used an acceptable approach.*
*8% failed to identify either the independent or the dependent variable.*

## Control of variables

Control in performing investigations refers to actual control, not intent. Therefore a pupil who has a poor measurement technique and is intending to control the volume of water will be considered to have failed to do so if the measurement has an error greater than 10%. Similarly an inability to use a thermometer correctly may result in the initial temperature not actually being controlled despite the pupil's intention to do this. In planning only stated intent can be assessed.

The variables considered in assessing the written plan were as follows: the fabric, the initial temperature of the water, the volume of water and the cooling of the cans. No obvious pattern was discernible in the choice of variables to be controlled, which was also the case in Category 6, in which 14% of the pupils controlled all the variables possible compared with 5% in Category 5. The 14% of the pupils who, in Category 6 investigations, actually controlled the variables underestimates those pupils who attempted to control. The numbers of pupils controlling two or more variables in Categories 6 and 5 were 80% and 51% respectively.

Five such planning questions were set to the sample of about 800 pupils and the results were examined to see if these pupils achieved a particular level of performance across different questions. This showed that the context of the questions had a marked effect; the same pupils achieved very different levels of performance on different questions.

## Proposing testable statements

The first step in the problem-solving chain is the perception of the problem. In order to assess this skill in a paper and pencil test an ambiguous or vague statement is given to the pupils and they are asked to rewrite it in a way that would make it scientifically testable. Of the two examples the first is in a science context, and the second in an everyday one.

**'Featherlight'** (n = 826)

A pupil says:

Feathers are lighter than lead.

Think of all the different things he might mean by this.

Some of these things could never be checked to find out if they are right or wrong, either because they are just someone's opinion, or else because they are not clear.

Choose one of the things he might mean that really <u>could be checked</u>, and write it down.

. . . . . . . . . . . . . . . . . . . . . . . . . . . . . . . .
. . . . . . . . . . . . . . . . . . . . . . . . . . . . . . . .
. . . . . . . . . . . . . . . . . . . . . . . . . . . . . . . .
. . . . . . . . . . . . . . . . . . . . . . . . . . . . . . . .

**'Brown Eggs'** (n = 830)

Some people say

It is better to eat brown eggs

Think of all the different things they might mean by this.

Some of these things could never be checked to find out if they are right or wrong, either because they are just someone's opinion, or else because they are not clear.

Choose one of the things they might mean that really <u>could be checked</u> and write it down.

. . . . . . . . . . . . . . . . . . . . . . . . . . . . . . . .
. . . . . . . . . . . . . . . . . . . . . . . . . . . . . . . .
. . . . . . . . . . . . . . . . . . . . . . . . . . . . . . . .
. . . . . . . . . . . . . . . . . . . . . . . . . . . . . . . .

| Pupil response | % of pupils |
|---|---|
| **Reformulation** | |
| A fixed volume of feathers is lighter than the same volume of lead | 8 |
| A fixed mass of feathers equals the same fixed mass of lead | 5 |
| A fixed mass of feathers is lighter than the same fixed mass of lead | 3 |
| **Investigation proposed – reformulation implicit** | |
| Drop a feather and a similar quantity of lead – lead will land first | 2 |
| Drop a feather and some lead – lead will land first | 9 |
| Feather placed in water would float, lead would sink | 2 |
| Same quantity of each – weigh them | 11 |
| Weigh them | 11 |
| **Explanation** | |
| In terms of properties of materials e.g. lead is heavier – feathers have no mass | 5 |
| In terms of the child's reality e.g. Birds are heavier than a pencil; it is no good for a bird to have feathers made of lead, etc. | 3 |
| Paraphrase original statement | 4 |
| Irrelevant/incomprehensible | 18 |
| No response | 12 |

| Pupil response | % of pupils |
|---|---|
| **Investigation proposed – reformulation implicit** | |
| Feed one group brown eggs, one group white eggs and see which is healthier | 3 |
| Check the freshness of brown eggs compared to white eggs | 8 |
| Check the bacteria in brown eggs compared to white eggs | 3 |
| See what brown eggs contain compared to white eggs | 5 |
| **Explanation** | |
| Brown eggs have more vitamins/minerals/calcium/protein than white eggs | 20 |
| Brown eggs are bigger/heavier/have more yolk than white eggs | 7 |
| Brown eggs have thicker shells | 1 |
| Brown eggs are fresher than white eggs | 2 |
| Brown eggs taste better/best/good to eat | 5 |
| Brown eggs are more nourishing/better for your health/good for you/richer | 14 |
| Brown eggs are laid by better/healthier/free-range chickens | 2 |
| Paraphrase original statement | 3 |
| Irrelevant/incomprehensible | 13 |
| No response | 15 |

The pupils' either rephrase and add controls and thus reformulate the statement, or they rephrase the statement as a test i.e. they go one step further in the chain. The other common category of response shown above is the one in which they try to explain the statement.

Their responses to these questions suggest the more 'scientific' the statement and the less accessible from the pupils' background knowledge and experience the more likely pupils are to behave as 'scientists'. In an everyday setting, as in 'Brown Eggs', an explanation rather than a reformulation seems the most appropriate response to the pupils. None of them gives a reformulation alone. About 35% of the pupils suggest an experimental procedure in their response for 'Featherlight' compared to 19% in 'Brown Eggs'. In 'Brown Eggs' 51% of the pupils attempt to explain the statement and do not enter the problem-solving chain. This compares with 8% in 'Featherlight'.

## Planning parts of an investigation

The assessment focuses on each aspect of planning in isolation.

During planning the pupils have to decide **what to vary**. Questions used to assess this particular aspect of planning present the pupil with a problem, an experimental set-up, and several possible important variables. For example, in 'Gases React' the pupil has to identify not only which variable to change, i.e. the independent variable, but also any which must be controlled.

**'Gases React'**

---

Some pupils were doing an investigation to find out how two different gases reacted with one another.

They knew that the temperature and the pressure might make a difference to the reaction.
One pupil thought that the substance of which the container was made would also matter.

To find out if he was right, they could do one of the things in the list below.

Put a tick in the box beside the one test you think they should do.

☐ A  Use different containers and different temperatures, but always keep the pressure the same.

☐ B  Use different containers, but always keep the temperature the same and the pressure the same.

☐ C  Use the same container all the time, but alter both the temperature and the pressure.

☐ D  Use different containers and different pressures, but always keep the temperature the same.

**Store distribution**
*n = 830*
*mean score = 52%*
*non-response = 3%*

**% of pupils**
*A  6*
*B  52 Key*
*C  24*
*D  13*

*M  (Multiple response) 2*

---

Many questions are concerned with **control of variables,** asking pupils either to state some necessary controls given both a problem and an experimental procedure to test it; or to select controls; or to criticise inadequate controls. The question 'London water' is an example where the pupils have to state three controls which are necessary.

### 'London water'

A group of pupils are comparing water from two different towns. They want to do a test to find out which kind of water lathers more easily with soap flakes.

If they want to make it a <u>fair</u> test they will have to make sure that some things in the test are the same for both kinds of water. Suggest <u>three</u> things that should be the same:-

(1) . . . . . . . . . . . . . . . . . . . . . . . . . . . . . .
. . . . . . . . . . . . . . . . . . . . . . . . . . . . . .
. . . . . . . . . . . . . . . . . . . . . . . . . . . . . .

(2) . . . . . . . . . . . . . . . . . . . . . . . . . . . . . .
. . . . . . . . . . . . . . . . . . . . . . . . . . . . . .
. . . . . . . . . . . . . . . . . . . . . . . . . . . . . .

(3) . . . . . . . . . . . . . . . . . . . . . . . . . . . . . .
. . . . . . . . . . . . . . . . . . . . . . . . . . . . . .

**Score distribution**
*n = 822*
*mean score = 70%*
*non-response = 8%*

| Pupil response | Score | 1 | 2 | 3 |
|---|---|---|---|---|
| Volume/mass/amount/level of water | 1 | 41 | 22 | 5 |
| Mass/number/volume/amount of soap flakes | 1 | 15 | 35 | 13 |
| Same brand of soap flakes | 1 | 19 | 8 | 7 |
| Same size of soap flakes | 1 | | | |
| Same water temperature/heat | 1 | 3 | 4 | 7 |
| Same amount of agitation/lathering/frothing up | 1 | | 1 | 8 |
| Same time of agitation/left for the same time | 1 | | 1 | 11 |
| Same source/kind of water | 1 | 3 | 3 | 4 |
| Same room condition | 0 | | | 1 |
| Bottles the same size | 0 | 3 | 5 | 6 |
| Same treatment for both | 0 | | | 2 |
| No interference with either e.g. nothing added | 0 | | 1 | 1 |
| Contradictory response - remove impurities/same purity/same content | 0 | 1 | 1 | 1 |
| Irrelevant | 0 | 5 | 5 | 6 |
| Incomprehensible | 0 | 2 | 2 | 3 |
| No response | 0 | 8 | 11 | 17 |

*The mean scores on the separate parts were 80%, 73% and 59% respectively.*

The fall in mean score across the three parts is understandable, as it becomes progressively more difficult for the pupil to generate another variable which needs to be controlled. Why this is so becomes a little clearer if one looks at the frequency of different responses. The pupils go for the most obvious variables first: volume of water, mass of soap flakes or brand of soap flakes. The rest of the possible variables are much less obvious and more knowledge-dependent, e.g. water temperature, agitation, source of water.

During planning, at the same time as deciding what to vary, the pupils must decide how to test the independent variable. How they define **what they should measure** indicates what they think the dependent variable is. The question 'Soils' is an example of one used in the survey. Here the pupil was given the problem and an experimental procedure to test it. The pupil had to decide what was wrong with the measurement procedure and offer an alternative. Only 28% of the pupils correctly evaluated the experimental procedure. Of those pupils only 17% could generate either completely or partially the correct dependent variable. The majority of the pupils' responses were to do with imagined or irrelevant practical details; for example 7% of the pupils who felt that Step 2 was wrong explained that it was pointless drying the soil as it would be made wet later on.

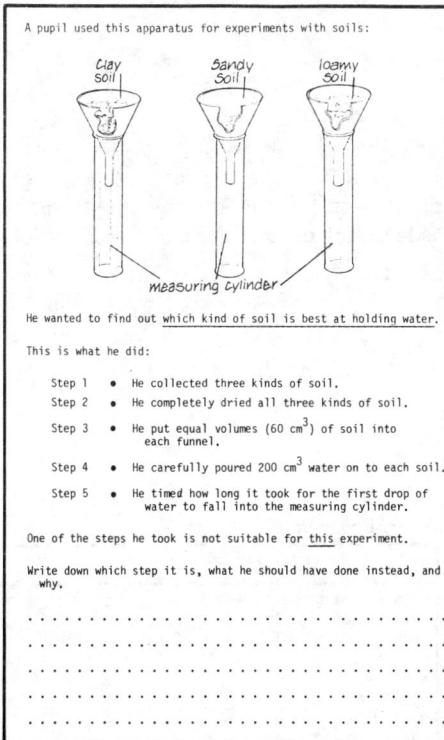

A pupil used this apparatus for experiments with soils:

Clay Soil    Sandy Soil    loamy Soil

measuring cylinder

He wanted to find out which kind of soil is best at holding water.

This is what he did:

Step 1  •  He collected three kinds of soil.
Step 2  •  He completely dried all three kinds of soil.
Step 3  •  He put equal volumes (60 cm$^3$) of soil into each funnel.
Step 4  •  He carefully poured 200 cm$^3$ water on to each soil.
Step 5  •  He timed how long it took for the first drop of water to fall into the measuring cylinder.

One of the steps he took is not suitable for this experiment.

Write down which step it is, what he should have done instead, and why.

. . . . . . . . . . . . . . . . . . . . . . . . . . . . . . . . . . . . . .
. . . . . . . . . . . . . . . . . . . . . . . . . . . . . . . . . . . . . .
. . . . . . . . . . . . . . . . . . . . . . . . . . . . . . . . . . . . . .
. . . . . . . . . . . . . . . . . . . . . . . . . . . . . . . . . . . . . .

**Score distribution**
*n = 826*
*mean score 17%*
*non-response 20%*

Pupil response

| Pupil response | | Score | % of pupils |
|---|---|---|---|
| Incorrect step: | Step 1 | 0 | 1 |
| | Step 2 | 0 | 27 |
| | Step 3 | 0 | 7 |
| | Step 4 | 0 | 17 |
| | Step 5 | 1 | 28 |
| No response | | | |
| Pupils' explanation for Step 5 (28% of the pupils) | | | |
| Measure the total volume/amount of water coming through at the end or in a certain time. The soil that lets the least water through holds it the best | | 2 | 8 |
| Measure the volume of water coming through until the end or for a certain time | | 1 | 9 |
| Water might drop at a slower pace through some soils | | 0 | 1 |
| Should evaporate the water and collect the vapour | | 0 | 1 |
| Should allow a set amount through e.g. 20 cm$^3$ | | 0 | 1 |
| Other response | | 0 | 4 |
| Irrelevant/incomprehensible | | 0 | 4 |

# Summary: PLANNING OF INVESTIGATIONS

When planning entire investigations pupils are less successful than when they have the opportunity to perform the investigation. In the non-practical context more pupils tend to choose non-scientific solutions and qualitative methods of judging the dependent variable. Pupil performance is low on control of variables and the selection of which variables to be controlled appears to be quite random.

In general the written form of a question as opposed to a practical one, and a written response compared to an action response, depress pupils' performance on control of variables. The results for the whole of 'Planning' indicate that if the effect of a variable can be understood only by applying scientific concepts, then such variables are generally not controlled.

Pupils' performance is relatively depressed on all questions which requires them to criticise an experimental set-up or a measurement procedure. Over 50% of the pupils in 1981 scored zero on these questions and this is consistent with previous survey results.

# INTERPRETATION AND APPLICATION
## (Category 4)

### INTERPRETATION

Two types of activity were assessed:
- **Interpreting presented information**
- **Distinguishing degrees of inference**

**Interpreting presented information** Questions in this sub-category present pupils with information in which some kind of relationship or pattern can be perceived. Some questions ask pupils to describe the relationship (that is, to generalise from the date), and others ask for predictions to be made on the basis of the perceived relationship. Two examples are shown.

**'Growth rings':** *describing a relationship*

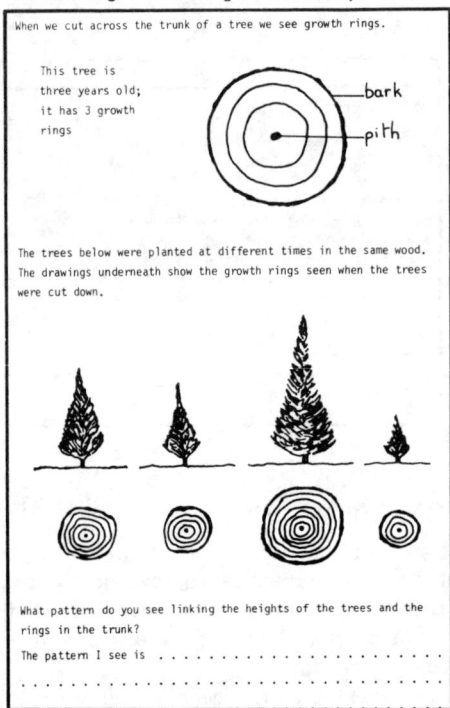

*How pupils responded*

| Type of response | Score | % of pupils (n=822) |
|---|---|---|
| The taller/bigger the tree, the more growth rings | 3 | 38 |
| Complete set of separate statements: tallest tree had most rings, etc. | 2 | 3 |
| Incomplete, e.g. the tallest tree had most rings. | 1 | 3 |
| Statements relating inappropriate variable: | | |
| height and age | | 13 |
| height to trunk | | 1 |
| age to trunk | | 1 |
| age to number of rings | | 2 |
| Single variables: | | |
| age or number of rings | | 6 |
| height only | | 1 |
| Mention of growth rate – spacing of rings | | 1 |
| Ambiguous – where 'big' could refer to height or diameter | | 1 |
| Visual pattern, e.g. contours, finger print, wood grain | | 6 |
| Other | | 8 |
| No response | | 8 |
| Meaningless/unreadable | | 8 |

This question was originally set for age 11 pupils. Results at both ages are shown below.

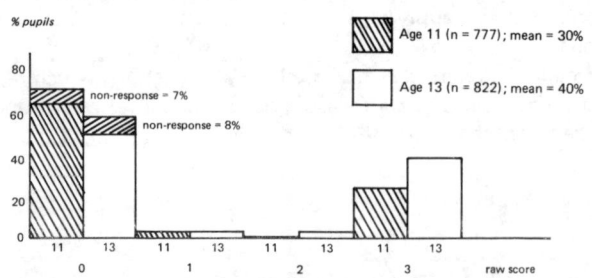

## 'Reflection': *selecting the 'odd man out'*

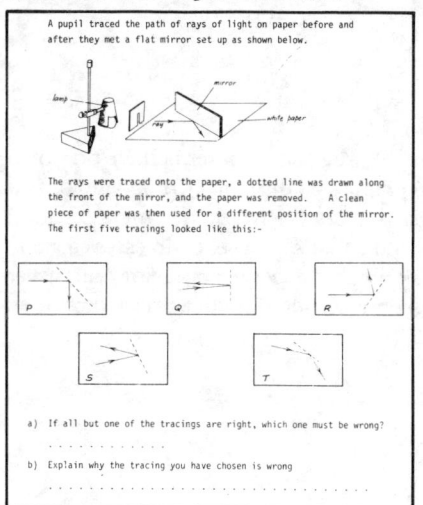

A pupil traced the path of rays of light on paper before and after they met a flat mirror set up as shown below.

The rays were traced onto the paper, a dotted line was drawn along the front of the mirror, and the paper was removed.    A clean piece of paper was then used for a different position of the mirror. The first five tracings looked like this:-

a) If all but one of the tracings are right, which one must be wrong?

.  .  .  .  .  .  .  .  .  .  .  .

b) Explain why the tracing you have chosen is wrong

.  .  .  .  .  .  .  .  .  .  .  .  .  .  .  .  .  .  .  .  .  .  .  .  .  .  .  .

n = 813
mean = 19%

|     |   | % pupils |                    |
|-----|---|----------|--------------------|
|     | P | 14       |                    |
|     | Q | 35       | (Note that Q is the |
|     | R | 14       | most popular choice) |
| Key | S | 19       |                    |
|     | T | 11       |                    |
|     | p | 7        |                    |

The pattern to be perceived in the data is the 'equal angles' law of reflection; it was not necessary for pupils already to have met this law.

**Distinguishing degrees of inference** Two examples of questions from this sub-category are illustrated below. The percentage of pupils selecting different statements is indicated on the questions.

### 'Candle'

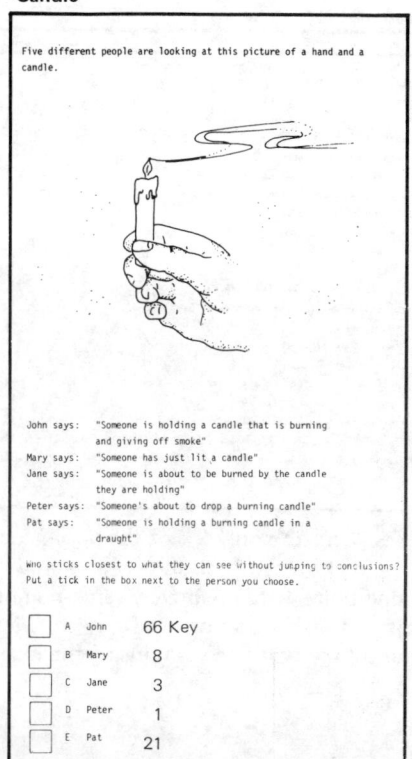

Five different people are looking at this picture of a hand and a candle.

John says:   "Someone is holding a candle that is burning and giving off smoke"
Mary says:   "Someone has just lit a candle"
Jane says:   "Someone is about to be burned by the candle they are holding"
Peter says: "Someone's about to drop a burning candle"
Pat says:    "Someone is holding a burning candle in a draught"

Who sticks closest to what they can see without jumping to conclusions? Put a tick in the box next to the person you choose.

| | | | |
|--|--|------|------|
| ☐ | A | John | 66 Key |
| ☐ | B | Mary | 8 |
| ☐ | C | Jane | 3 |
| ☐ | D | Peter | 1 |
| ☐ | E | Pat | 21 |

### 'Vase of flowers'

Five different people are looking at this picture.

John says:     The radiator has been turned up too high.
Mary says:     The vase has got no water in it.
Jane says:     The flower stems are too weak.
Peter says:    The vase has got drooping flowers in it.
Pat says:      The people must be away in this house.

Who sticks closest to what they can see without jumping to conclusions? Put a tick in the box next to the person you choose.

| | | | |
|--|--|------|------|
| ☐ | A | John | 26 |
| ☐ | b | Mary | 15 |
| ☐ | C | Jane | 2 |
| ☐ | D | Peter | 43 Key |
| ☐ | E | Pat | 10 |

# APPLICATION

Two types of activity were assessed:
- Applying Science concepts
- Generating alternative hypotheses

**Applying science concepts:** In these questions a situation is described in one of a variety of ways (using words, tables of figures, graphs, or diagrams, for example) and pupils have to recall and use science concepts they are likely to have met in school science lessons. They may have to explain how the situation arose, or to assess a statement about it, or to predict what will happen if circumstances change. They are never asked simply to state a learned fact, or to explain a standard application which they may have learned by heart.

**'Stream':** *giving an explanation*

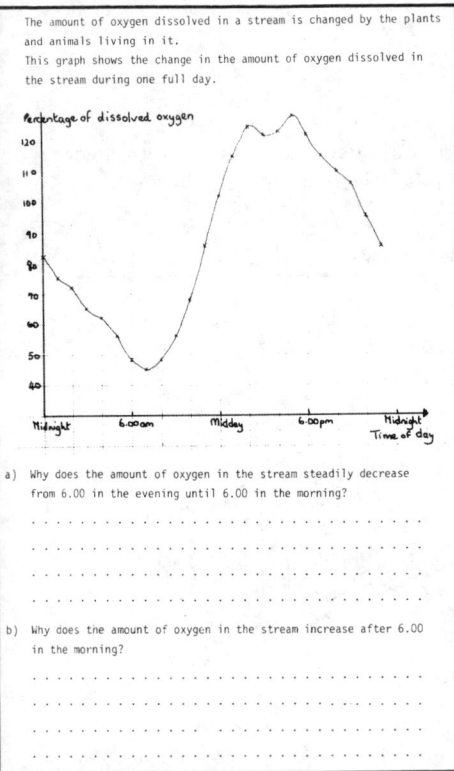

The amount of oxygen dissolved in a stream is changed by the plants and animals living in it.
This graph shows the change in the amount of oxygen dissolved in the stream during one full day.

a) Why does the amount of oxygen in the stream steadily decrease from 6.00 in the evening until 6.00 in the morning?

. . . . . . . . . . . . . . . . . . . . . . . . . . . . . . .

. . . . . . . . . . . . . . . . . . . . . . . . . . . . . . .

. . . . . . . . . . . . . . . . . . . . . . . . . . . . . . .

. . . . . . . . . . . . . . . . . . . . . . . . . . . . . . .

b) Why does the amount of oxygen in the stream increase after 6.00 in the morning?

. . . . . . . . . . . . . . . . . . . . . . . . . . . . . . .

. . . . . . . . . . . . . . . . . . . . . . . . . . . . . . .

. . . . . . . . . . . . . . . . . . . . . . . . . . . . . . .

. . . . . . . . . . . . . . . . . . . . . . . . . . . . . . .

## Responses for 'Stream'

| Type of response | Score | % pupils |
|---|---|---|
| **part a) (Why does oxygen decrease overnight?)** | | |
| Oxygen is used in respiration of animals and plants, none produced by photosynthesis | 2 | 2 |
| Less sunlight for photosynthesis | 1 | 3 |
| Sun goes down/it gets dark | 1 | 6 |
| Plants use oxygen at night | 1 | 4 |
| More creatures arrive and use up oxygen | 0 | 5 |
| Animals are more active and use up oxygen | 0 | 7 |
| Animals and plants use it at this time | 0 | 4 |
| Animals/plants go to sleep and don't need oxygen | 0 | 25 |
| No response | 0 | 15 |
| Others | 0 | 29 |
| **part b) (Why does oxygen increase during day?)** | | |
| Oxygen used in respiration but more is produced by photosynthesis | 2 | 1 |
| More sunlight for photosynthesis | 1 | 4 |
| Sun comes up/gets light | 1 | 6 |
| Plants do not use oxygen in day time | 1 | 7 |
| Fewer animals in stream – go away for day | 0 | 3 |
| Animals less active so don't use oxygen up | 0 | 5 |
| Animals wake up and need more oxygen | 0 | 23 |
| Animals and plants do not use it | 0 | 1 |
| No response | 0 | 18 |
| Other | 0 | 32 |

Few pupils gave more than a partial answer to 'Stream'; only 2% of the pupils took into account the balancing effect of respiration and photosynthesis in part a), and only 1% in part b). But perhaps the most interesting point is the relatively large number (25%) who considered, in part a), that animals and plants do not need oxygen when they are 'asleep' and that as a consequence the supply will decrease during the night.

A similar proportion of pupils (23%) seemed to think that animals, waking up, need more oxygen and that the supply will, therefore, increase during the day. The mean score was 10% and the non-response rate 13%.

**'Torch'**: *giving an explanation*

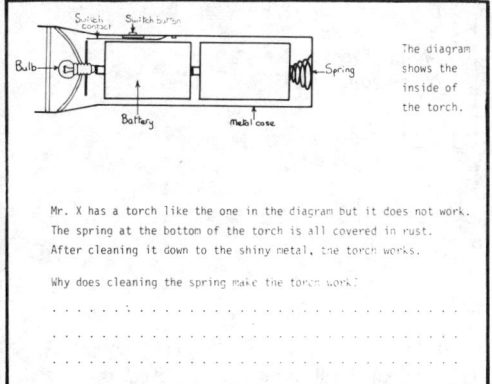

The diagram shows the inside of the torch.

Mr. X has a torch like the one in the diagram but it does not work. The spring at the bottom of the torch is all covered in rust. After cleaning it down to the shiny metal, the torch works.

Why does cleaning the spring make the torch work?

. . . . . . . . . . . . . . . . . . . . . . . . . . . . . . . .

. . . . . . . . . . . . . . . . . . . . . . . . . . . . . . . .

. . . . . . . . . . . . . . . . . . . . . . . . . . . . . . . .

*The mean score was 21%.*
*4% of the pupils failed to respond.*
*Only 2% of the pupils obtained all 3 marks.*

**Mark Scheme — 'Torch'**

| | |
|---|---|
| Complete circuit needed for torch to work | 1 |
| Spring forms part of the circuit | 1 |
| Rust is a non-conductor so no circuit | 1 |
| Total | ③ |

% pupils

*Bar chart showing % pupils against Score (0, 1, 2, 3). Bar at 0 is about 55% with hatched portion near top; bar at 1 is about 23%; bar at 2 is about 15%; bar at 3 is about 3%.*

**'Salty water'**: *selecting a prediction and justifying the choice*

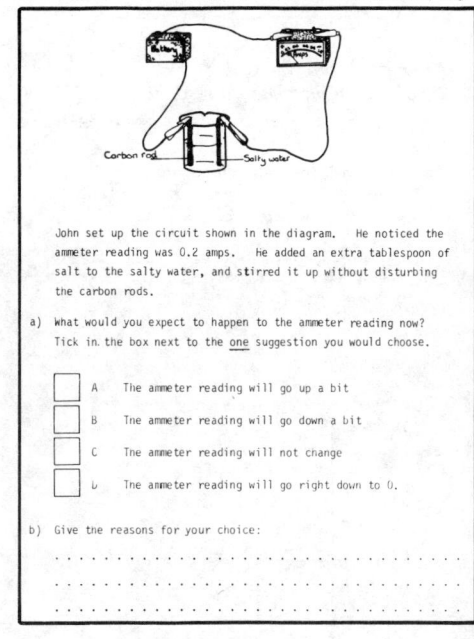

John set up the circuit shown in the diagram. He noticed the ammeter reading was 0.2 amps. He added an extra tablespoon of salt to the salty water, and stirred it up without disturbing the carbon rods.

a) What would you expect to happen to the ammeter reading now? Tick in the box next to the one suggestion you would choose.

☐ A   The ammeter reading will go up a bit

☐ B   The ammeter reading will go down a bit

☐ C   The ammeter reading will not change

☐ D   The ammeter reading will go right down to 0.

b) Give the reasons for your choice:

. . . . . . . . . . . . . . . . . . . . . . . . . . . . . . . .

. . . . . . . . . . . . . . . . . . . . . . . . . . . . . . . .

. . . . . . . . . . . . . . . . . . . . . . . . . . . . . . . .

**Mark Scheme**

| | |
|---|---|
| a) Key A | 1 |
| b) More salty/concentrated or stronger the solution the bigger/faster the current or the better the conduction | 2 |
| Total | ③ |

*In part a) 60% of the pupils selected the correct answer.*
*In part b) 10% of them gained the full 2 marks and a further 14% gained 1 mark.*
*11% of the pupils made no attempt to give the explanation.*

**'pH'** : *applying science concepts to select a prediction*

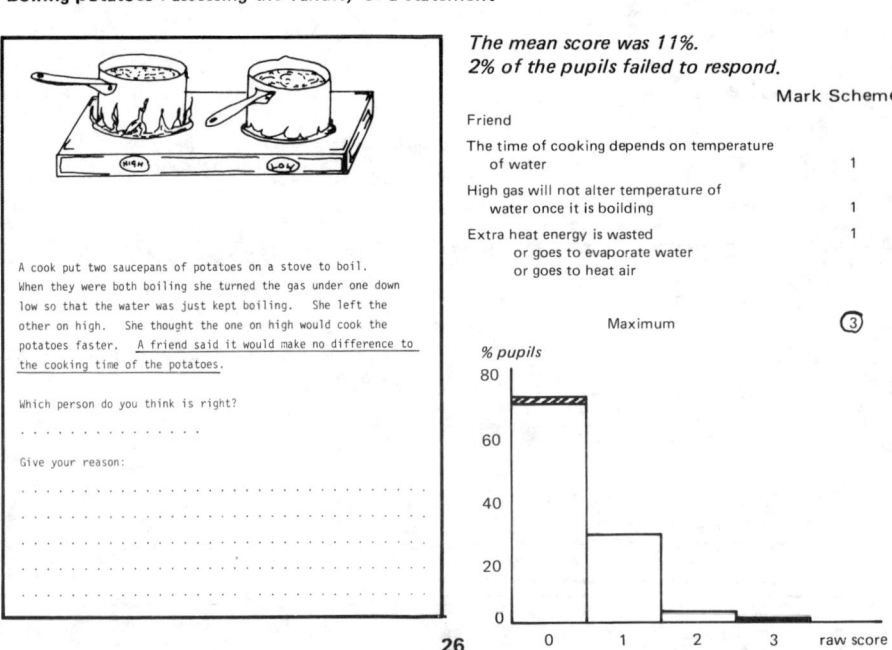

A pupil was trying to change the pH (acidity) of 10 cm³ of nitric acid by adding something to it.

Which <u>one</u> of the following suggested additions would <u>not</u> alter the acidity?   Tick in the box next to the suggestion you choose.

- [ ] A   5 g of calcium metal
- [ ] B   10 cm³ of the same nitric acid
- [ ] C   5 g of magnesium metal
- [ ] D   10 cm³ of alkali
- [ ] E   10 cm³ of pure water

|   | % pupils |
|---|---|
| A | 4 |
| B (Key) | 34 |
| C | 7 |
| D | 18 |
| E | 28 |

*Multiple response = 5%*
*No response = 34%*

Although the correct response (more of the same nitric acid) gained the highest proportion of responses (34%), many pupils were understandably attracted by the idea of adding pure water. It is more difficult to see why 18% should have chosen the addition of alkali, unless they missed the negative in the stem of the question in spite of the word *'not'* being underlined; perhaps they thought in terms of adding alkali to acid from force of habit.

**'Boiling potatoes'**: *assessing the validity of a statement*

A cook put two saucepans of potatoes on a stove to boil. When they were both boiling she turned the gas under one down low so that the water was just kept boiling.   She left the other on high.   She thought the one on high would cook the potatoes faster.   <u>A friend said it would make no difference to the cooking time of the potatoes.</u>

Which person do you think is right?

. . . . . . . . . . . . . . .

Give your reason:

. . . . . . . . . . . . . . . . . . . . . . . . . . . . . .
. . . . . . . . . . . . . . . . . . . . . . . . . . . . . .
. . . . . . . . . . . . . . . . . . . . . . . . . . . . . .
. . . . . . . . . . . . . . . . . . . . . . . . . . . . . .
. . . . . . . . . . . . . . . . . . . . . . . . . . . . . .
. . . . . . . . . . . . . . . . . . . . . . . . . . . . . .

*The mean score was 11%.*
*2% of the pupils failed to respond.*

**Mark Scheme**

Friend

| | |
|---|---|
| The time of cooking depends on temperature of water | 1 |
| High gas will not alter temperature of water once it is boiling | 1 |
| Extra heat energy is wasted or goes to evaporate water or goes to heat air | 1 |
| Maximum | ③ |

**Generating alternative hypotheses** Pupils are asked to account for a situation in several different ways. The variety of responses to the example 'Fountain' is shown in the table below. The percentages of pupils giving a particular response to the three parts of the question are shown in columns 1, 2 and 3.

'Fountain'

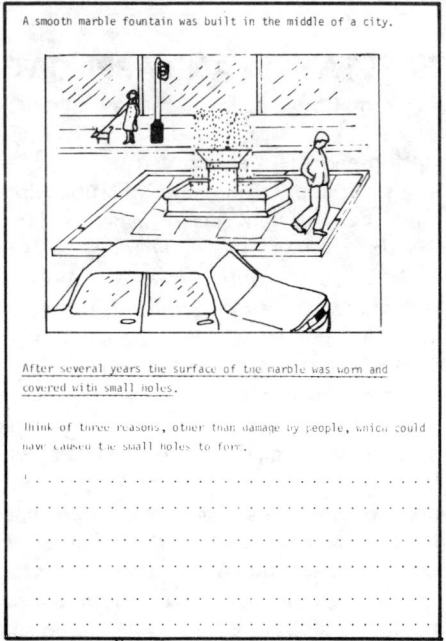

A smooth marble fountain was built in the middle of a city.

After several years the surface of the marble was worn and covered with small holes.

Think of three reasons, other than damage by people, which could have caused the small holes to form.

*Responses for 'Fountain'*

| Response | Score | % of pupils: | | | Overall Incidence /300 |
|---|---|---|---|---|---|
| | | (i) | (ii) | (iii) | |
| Water in fountain reacts with/ wears away marble | 1 | 39 | 16 | 8 | 63 |
| Weathering by rain and wind | 1 | 17 | 16 | 12 | 45 |
| Effect of pollutants in the air | 1 | 11 | 10 | 7 | 28 |
| Caused by insects/plants/fungi | 1 | 2 | 5 | 3 | 10 |
| animals | 1 | 2 | 5 | 4 | 11 |
| birds | 1 | 1 | 1 | 1 | 3 |
| Stones thrown up by cars chip the marble | 1 | 0 | 1 | 2 | 3 |
| Other possible causes | 1 | 2 | 2 | 2 | 6 |
| People (ruled out in the question) | 0 | 7 | 8 | 7 | 22 |
| Holes already there | 0 | 1 | 1 | 1 | 3 |
| Other irrelevant suggestions | 0 | 15 | 21 | 22 | 58 |
| No response | 0 | 5 | 12 | 29 | 46 |
| Mean score for separate parts | | 73% | 52% | 36% | |

The most popular suggestion overall was that the action of the water in the fountain was responsible for the formation of the holes in the marble. This was also the one most likely to be put first. In spite of the phrase 'other than damage by people' which appears in the question, human action was suggested as the cause by 7% of the pupils. Perhaps this reflects public concern over vandalism, as well as lack of attention to instructions.

## Summary: INTERPRETATION AND APPLICATION

**Interpreting presented information** has an overall mean score of about 40%, many questions failing in the 25% to 50% range. Pupils' difficulties seem to be associated with the type of relationship and the manner of its presentation as well as with other demands. In general, pupils can apply a perceived relationship more successfully than they can describe it; and their level of performance is often depressed when the question is set in a scientific context. This may be because pupils assume that just because a question is set in a science *context* it will require understanding of science *concepts;* they are thus reluctant to do simply what the question demands: to interpret the information presented. A contrast to this is provided by the type of question in which pupils are asked to select the statement which keeps most closely to the presented information. In this case, the more everyday the circumstances of the event described, the more likely the pupils are to jump to unwarranted conclusions. Thus in **Distinguishing degrees of inference** a science context tends to *enhance* performance.

When science concepts have to be **applied** in order to make sense of information, the level of performance drops to less than 30%. Where questions involve a combination of novel context, the need for understanding of science concepts and a sophisticated process such as the assessment of a presented hypothesis, scores are very low. In each concept region, the effect of question type appears to override concept considerations, within the limits set by the framework; for example, pupils have higher mean scores for questions requiring generation of predictions, on average, than for those calling for explanations or assessment of alternative hypotheses. In general, pupils generate a great variety of **alternative hypotheses** when asked to do so; however they tend to be more reluctant to make suggestions when the subject matter of a question appears to them to be scientific than in other cases.

# OBSERVATION: (Category 3)

Observation is tested by a variety of practical tasks; many of them demand the examination of real objects or events or photographs. In some cases pupils have to make hypotheses or generalisations. In general, the resources chosen are such that pupils have to select what to observe for themselves. Questions such as "What number is the pointer indicating on the scale of the meter?" are regarded as focusing on **Using measuring instruments** (Category 2) rather than on observation; those like "Is the object black or white?" do not figure in the assessment at all, except as a means of checking by written response whether certain actions have been taken. There are no sub-divisions in this category.

Tests were administered to groups of nine pupils by a visiting trained science teacher with help from a science teacher from the survey school. Pupils rotated round a circus of nine stations, being allowed eight minutes at each station. The number of questions at a station varied; each example quoted occupied a station. The visiting tester, in addition to organising the circus, was responsible for marking the completed scripts using a previously tested and agreed scheme.

## Making and interpreting observations

**'Tracks':** *stating similarities and difference*

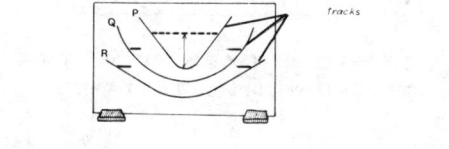

*n = 975*
*mean score = 28%*

You have been given a ball-bearing and tracks of three different shapes labelled P, Q and R.
Your job is to find a number of differences and similarities between the movement of the ball-bearing on the different tracks. Hold the ball-bearing on each track in turn at the mark and let it go.

a) Write down three ways in which the movement of the ball-bearing on the three tracks is different. Take each track in turn and say how it is different.

| | P | Q | R |
|---|---|---|---|
| 1. | | | |
| 2. | | | |
| 3. | | | |

b) Write down three ways in which the movement of the ball-bearing is same on each track.

1. ....................
2. ....................
3. ....................

**How pupils responded**

| Part a) Differences stated (Mean score 33%) | Frequency of statement % |
|---|---|
| Path length | 22 |
| 'Speed' of swing | 23 |
| Duration of swings | 24 |
| Total number of swings | 2 |
| Noise | 3 |
| Irrelevant | 25 |
| No response | 1 |
| **Part b) Similarities stated (Mean score 17%)** | |
| Swing backwards and forwards | 19 |
| All finally stop at the centre | 18 |
| Swing a number of times | 1 |
| The length of swing (amplitude) decreases | 4 |
| The balls roll | 3 |
| Noise | 1 |
| Irrelevant | 31 |
| No response | 23 |

**'Tadpoles'**: *stating similarities and differences*

*n = 969*
*mean score = 40%*

*How pupils responded*

You have been given a hand lens and five tubes labelled P, Q, R, S and T containing animals.

You can pick up each tube and turn it around.

a) Look carefully at the animals in S and R.   One animal is bigger than the other.   Apart from this write down <u>four</u> things that are <u>different</u> about them.

1. . . . . . . . . . . . . . . . . . . . . . . . . . . . . . .

2. . . . . . . . . . . . . . . . . . . . . . . . . . . . . . .

3. . . . . . . . . . . . . . . . . . . . . . . . . . . . . . .

4. . . . . . . . . . . . . . . . . . . . . . . . . . . . . . .

b) Animal S is bigger than animal Q.   Apart from this what is the main difference between the head of the animal in S and the head of Q?

. . . . . . . . . . . . . . . . . . . . . . . . . . . . . . .

c) Look carefully at P and T.   Write down five ways in which these animals are alike.

1. . . . . . . . . . . . . . . . . . . . . . . . . . . . . . .

2. . . . . . . . . . . . . . . . . . . . . . . . . . . . . . .

3. . . . . . . . . . . . . . . . . . . . . . . . . . . . . . .

4. . . . . . . . . . . . . . . . . . . . . . . . . . . . . . .

5. . . . . . . . . . . . . . . . . . . . . . . . . . . . . . .

a) Mean score 33%.

| Number of differences stated: | % of pupils |
|---|---|
| 4 | 18 |
| 3 | 34 |
| 2 | 29 |
| 1 | 18 |
| 0 (non-response) | 1 |

b) Mean score 51%.

| Number of similarities stated: | % of pupils |
|---|---|
| 5 | 29 |
| 4 | 20 |
| 3 | 15 |
| 2 | 13 |
| 1 | 15 |
| 0 (non-response) | 8 |

**Comment** The five specimen tubes to which the question refers each contain a preserved tadpole. The tadpoles are at different stages of development, as indicated in the diagram.

The lower overall mean score for 'Tracks' (28%) may be due to the fact that the comparisons to be made are of a less concrete form. They are also separated in time, whereas those in 'Tadpoles' can be made from almost instantaneous observation;

You have been given three solids jn test tubes labelled P, Q and R

a) Take test tube P and look carefully at the solid.
   Place the test tube in a holder and heat it for about a minute.
   Write down the changes you notice in the test tube.

   . . . . . . . . . . . . . . . . . . . . . . . . . . . . .

   . . . . . . . . . . . . . . . . . . . . . . . . . . . . .

   . . . . . . . . . . . . . . . . . . . . . . . . . . . . .

   Repeat these instructions for test tubes Q and R:

b) Test tube Q

   Changes during heating in the test tube: . . . . . . . . . . . .

   . . . . . . . . . . . . . . . . . . . . . . . . . . . . .

   . . . . . . . . . . . . . . . . . . . . . . . . . . . . .

c) Test tube R

   Changes during heating in the test tube: . . . . . . . . . . . .

   . . . . . . . . . . . . . . . . . . . . . . . . . . . . .

   . . . . . . . . . . . . . . . . . . . . . . . . . . . . .

d) What do the changes for P, Q and R have in common with each other?

   . . . . . . . . . . . . . . . . . . . . . . . . . . . . .

   . . . . . . . . . . . . . . . . . . . . . . . . . . . . .

   . . . . . . . . . . . . . . . . . . . . . . . . . . . . .

   . . . . . . . . . . . . . . . . . . . . . . . . . . . . .

*Place all dirty test tubes in the container provided.*

**Comment**

*The pupils are given three test tubes containing coloured crystals as follows:*
*test tube P — nickel II sulphate*
*test tube Q — copper II sulphate*
*test tube R — iron II sulphate*

| Response | | % of pupils | | | |
|---|---|---|---|---|---|
| | | a) | b) | c) | d) |
| Responses – change in colour only | | 32 | 51 | 50 | 29 |
| – change in form only | | 2 | 1 | – | 1 |
| –(water)vapour given off | | 7 | 2 | 1 | 4 |
| Change in colour and form | | 14 | 6 | 7 | 4 |
| Change in colour and (water) vapour given off | | 28 | 29 | 24 | 17 |
| Cnange in colour, form and (water) vapour given off | | 5 | 2 | 1 | 1 |
| Irrelevant response | | 8 | 4 | 3 | 6 |
| No response | | 1 | 4 | 14 | 38 |
| Mean scores (%) | | 50 | 46 | 40 | 27 |

The change in mean score across parts a), b) and c) of 'Heating' was due to more pupils focusing on the colour changes alone in the later chemical events. This could be a consequence of the pupils already having formed the generalisation that colour changes were in common. Perhaps as a result of this focused observation other features were no longer noticed. The sharp decrease in mean score in part d) was due largely to the increase in non-responses. The most common observations were the colour changes followed by the loss of (water) vapour. Many pupils described the colour changes in great details or observed irrelevancies like black on the outside of the test tube. These pupils did not 'notice' the change in form as being significant although it was quite marked in each case.

## Summary: OBSERVATION

The pupils' pre-existing view of the world can adversely affect their performance in two ways. Firstly if the situation presented does not fit into any accepted framework then what to observe as relevant is unknown. Secondly if the situation to be observed accords with their prior experience of the world and the pupils feel confident about their experience they will often respond by giving information which is non-observable;

for example, that animals with legs jump, that unseen substances are shiny — both of which involve recall rather than observation. Essentially these pupils are interpreting their observations when the questions only require them to report their observations.

The surveys have shown that pupils are successful in making and using observations across a variety of contexts. The success is indicated by the fact that the overall mean score of 49% results from about half the questions used having mean scores between 40% and 60%, with some over 75%. Some questions had mean scores as low as 20%: these were often those in which pupils were not specifically directed as to what to observe.

# USE OF APPARATUS AND MEASURING INSTRUMENTS (Category 2)

Some distinctions between questions used to test this category and those used to test **Observation** have already been made. In addition, there are differences in the way responses are marked. In observation, the pupils' written words are the basis for assessment. In **Using apparatus and measuring instruments** it may also be their *actions,* (recorded on a checklist) or even a *product*, such as a measured quantity of sand.

Three kinds of activity were assessed:
  - **Using measuring instruments**
  - **Estimating quantities**
  - **Following instructions for practical work**

Tests were administered in the same way as those for Category 3; pupils rotated around nine stations set out as a 'circus'. Their movement, at eight-minute intervals, was organised by a visiting tester; and a science teacher from the survey school took charge of the station at which pupils' actions were recorded on a checklist.

**Using measuring instruments** In this sub-category questions require pupils to read scales, use appropriate units, and to use measuring instruments and other laboratory apparatus in clearly defined situations. There is no element of 'designing an investigation' here, and as has been indicated, any observation needed is at a very low level.

## 'Scale readings'

For each part of this question there are 2 marks for the value and 1 for the unit, giving a possible total of 3. Note, however, that the correct unit does not score a mark if the value given lies outside the wider of the two ranges specified.

Whether or not the pupils recorded a reasonable numerical value appeared to depend on several factors:
  - the shape of the scale;
  - The value of the smallest scale division (for the ammeter this was 0.02 A, the instrument having the commonly used 0—1 A range);
  - The need to take two readings, as in the stopclock and manometer;
  - the need to find the difference between two readings, as in the manometer.

Station ◯

> Seven different measuring instruments have been
> set up all ready for you to look at.   Do not
> change them in any way.
> Answer as many of the questions below as you can.
> Don't forget to say what units you are using.
> The first question has been answered to show
> what we mean.

How wide is this booklet?    | 21 cm |

a) How much water is there in the measuring cylinder?    | |

b) How big is the force with which the rubber band pulls on the hook?    | |

c) What is the pressure of the gas at this gas-point?    | |

d) What is the temperature of the water in the flask?    | |

e) How long had the stop-clock been running before it was stopped?    | |

f) Press the push-button to switch on the current. Read the current through the circuit on the ammeter.    | |

g) Press the push-button again.   Read the voltmeter.    | |

*In each case, the 'close' range was the true value +/- the smallest scale division and the 'wide range' +/- two such divisions.*

| | | | | | Mean Score | Zero Score |
|---|---|---|---|---|---|---|
| | Value | Close range 2 | Wide range (1) | Units 1 | % | % pupils |
| a) measuring cylinder | 42 | ±1 | ±2 | $cm^3$ or cubic centimeter (not cc) | 57 | 24 |
| b) forcemeter | 27 | ±1 | ±2 | N or newton (allow Newtons) | 57 | 27 |
| c) manometer | ? | ±1 | ±2 | cm of water (not just cm) | 10 | 86 |
| d) thermometer | 35 | ±1 | ±2 | $^0$C or $^0$Centigrade or $^0$Celius (or degrees in full) | 80 | 7 |
| e) stop-clock | 2m 23s | ±1 | ±2 | m and s or minutes and seconds (allow min. and sec.) | 57 | 35 |
| f) ammeter | 0.38 | ±0.02 | ±0.04 | A or amp or ampere | 17 | 72 |
| g) voltmeter | 3.8 | ±0.1 | ±0.2 | V or volt | 63 | 19 |

It might have been expected that degree of familiarity would have accounted for differences in performance; but even though the ammeter is more commonly used than the voltmeter at this age, and the stopclock more familiar than either, the mean score was much higher for the voltmeter (67%) than for either stopclock (50%) or ammeter (30%).

Other questions involved the actual *use* of apparatus. Over 70% of the pupils could successfully use both hand lens and microscope to identify four randomly chosen capital letters which were too small to see with the naked eye, and a similar proportion a 100 cm$^3$ x 1 cm$^3$ measuring cylinder to measure out water to within 1 cm$^3$ of the specified volume. However, when it came to using a single scale lever arm balance they were not nearly so successful; in fact nearly 70% failed to get within 6 grams — that is within +/- 3 of the smallest marked scale divisions — when weighing out sand in a metal container.

**Estimating quantities** Two contrasting kinds of question are used. In one pupils are confronted with pre-determined amounts of material and asked to estimate the quantity present — the volume of liquid or the length of wire, for example; in the other, they are given a considerable surplus of material and asked to separate out a specified quantity without benefit of measuring instruments.

In 'Roundabout', pupils selected the amount of material they considered to match the specification, and left it behind in a lettered container for the administrator to check after the testing was over. Two of the questions required estimates of length, one of mass, and one of volume.

**'Roundabout'**

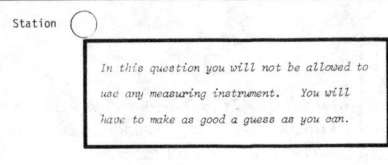

In this question you will not be allowed to use any measuring instrument. You will have to make as good a guess as you can.

a)  Tear off a length of ticker-tape 75 cm long.
    Put it in the envelope with your letter on it.

b)  Take a plastic bag with your letter on it.
    Put 100 grams of sand in it. Seal the bag with a
    twist-tie and 'post' it in the box provided.

c)  Take the beaker with your letter on it.
    Put 100 cm$^3$ of water into it from the tap.
    Put the beaker and water carefully into the box
    provided.

d)  Draw a line 11 cm long in the space below.
    You may use the 'straight edge' to help you.

$n = 1,032$

How well pupils estimated quantities (% of pupils per range)

|  | R a n g e | | | | |
| --- | --- | --- | --- | --- | --- |
|  | Close (+/-10%) | Wide (+/-20%) | Very wide (+/-50%) | Beyond +/-50% | Mean scores % |
| Tape (75cm) | 22 | 27 | 24 | 27 | 47 |
| Sand (100g) | 11 | 10 | 31 | 48 | 27 |
| Water (100cm$^3$) | 22 | 19 | 48 | 11 | 50 |
| Line (11 cm) | 37 | 25 | 25 | 13 | 63 |
| Score: | 3 | 2 | 1 | 0 | |

It seems curious that almost half the pupils failed to get within 50% of the mass of sand required, while only a tenth of them were as wide of the mark when considering volume of water. It is possible that this effect is due to more practice in one than the other — most pupils are asked to measure out a given volume of liquid more often than to measure out a given mass of material. Perhaps performance is also affected by the fact that estimating volume depends on sight, while estimating mass depends on kinesthetic sense — a consciousness of muscular effort. This would tally with results already noted: 13 year old pupils tend not to use senses other than sight in making observations unless specifically asked to do so.

In a set of questions in which pupils were asked simply to *state* estimates, they were less successful. Only 30% gave the length of a 15 cm piece of wire as between 13.5 and 16.5 cm — and this proved the easiest of the set. Over 10% of the pupils estimated lukewarm water (at $40^0$C) to be either below $10^0$C or above $80^0$C.

**Following instructions for practical work** Instructions of various kinds are used, both written and diagrammatic. In some cases they can be followed only by pupils who have learned the necessary techniques in a science laboratory, but in others they are complete in themselves.

**'Solid P'**: *following instructions using standard techniques*

Station ◯

You have been given a white solid labelled P and you are to find out about it by following these instructions. Use the apparatus that seems best for the job.

1. Gently heat a small amount of the solid P in a test tube. Hold a piece of dry cobalt chloride paper inside the mouth of the tube. Watch carefully.

2. Take some more of the solid and put it into a clean tube. Add a few drops of dilute acid to it. Watch what happens.

3. Do this again and test to see if the gas carbon dioxide is given off.

What kind of substance do you think this might be?

. . . . . . . . . . . . . . . . . . . . . . . . . . .

*The pupil's responses to this question were observed. A member of the science staff from the survey school was asked to fill in a separate checklist for each pupil, and to service the station by setting the apparatus out in a standard way and replacing used items between session. Because there was no necessity to depend on the pupil's written record it was possible to take account of skills that otherwise could not have been tested.*

Differences in skills observed in 'Solid P'

|  | % of pupils |
|---|---|
| Heated solid correctly | 19 |
| Used tweezers rather than fingers to handle cobalt chloride paper | 17 |
| Attempted the $CO_2$ test | 44 |
| Tested for $CO_2$ using: glass rod | 2 |
| pipette | 3 |
| tipping | 4 |
| Tested for $CO_2$ by adding lime water to crystals | 10 |

The mean score was 47% and the non-response rate less than 1%. The distribution of scores was such that more pupils achieved a score in the middle of the possible range than at either end.

In other questions, pupils followed instructions for tasks ranging from the construction of a kaleidoscrope to the identification of an unknown liquid. Mean scores were generally between 45% and 80%, but in questions which involved the setting up of a circuit from a conventional diagram it was down to less than 30%.

## Summary: USE OF APPARATUS AND MEASURING INSTRUMENTS

Pupils are poor at reading some of the measuring instruments tested. The difficulties sometimes seem to be associated with the scale, as in the case of the 0 - 1 A ammeter which has 0.02 A as the smallest scale division. But sometimes they appear to be due to lack of familiarity: The manometer used has a very simple scale, with 1 cm as the smallest scale division — but pupils who have not used it before may not realize that it is a *difference* in levels which indicates the excess pressure. They are much better at using measuring cylinders than lever arm balances; perhaps this is because they are more readily available in schools.

Correspondingly, pupils are much better at estimating a volume of liquid than a mass of any kind; but they are not really good at estimating any physical quantity except a length under 30 cm.

They are quite capable of following clear instructions, which may include diagrams, up to the point where they have to recall science specific techniques or symbols.

Pupils were reported as enjoying the practical tests, both in this category and others, and as working remarkably quickly and without undue fuss.

# USE OF GRAPHICAL AND SYMBOLIC REPRESENTATION (Category 1)

Lack of competence in the use of various ways of presenting information may be almost as much of a handicap, for anyone needing to communicate in a scientific context, as an inability to read and write well.

Three types of activity were assessed:
- **Reading information from graphs, tables and charts**
- **Expressing information as graphs, tables and charts**
- **Using scientific symbols and conventions**

**Representing information as graphs, tables and charts** Questions had similar mean scores, and responses were similarly affected by fractions and complexity. Provided the axes were drawn, scaled and labelled, many pupils could draw line graphs even when the context of the question was not all that familiar, as in 'Disappearance'. Putting suitable linear scales on the axes of graphs appeared to be the major difficulty. A popular alternative was to space out evenly along the axis any values that happened to be given.

**'Disappearance'**

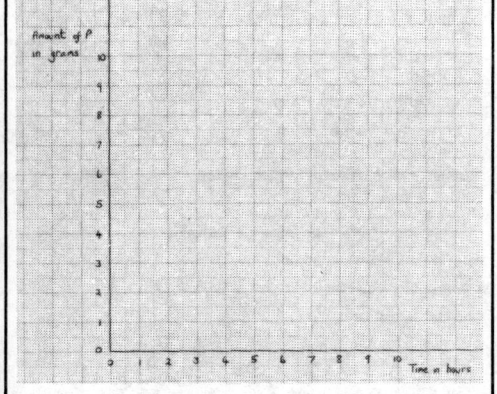

The table shows how a substance 'P' gradually disappears during a chemical reaction:

| Time in hours | 0 | 2 | 4 | 6 | 8 | 10 |
|---|---|---|---|---|---|---|
| Amount of 'P' left in grams | 8.0 | 6.4 | 5.3 | 4.4 | 3.7 | 3.2 |

Plot a line graph to illustrate these figures.

*The mean score was 60%. 20% of the pupils failed to score; but 46% scored full marks. The variety of styles of response is shown in the table below.*

| 'Disappearance': Style of graph/chart drawn | % of pupils |
|---|---|
| Points joined by straight lines | 48 |
| Curve through points | 14 |
| Vertical bars separated | 6 |
| Isolated crosses (or points) | 4 |
| Vertical bars together | 3 |
| Crosses and vertical lines | 3 |
| Crosses joined to both axes | 1 |
| Best straight line | 1 |
| Other | 11 |
| No response | 9 |

37

'Ammeter': *proposing names for components*

**Using scientific symbols and conventions** The symbols used at age 13 were confined to those used in circuit diagrams and to section drawings of common general apparatus such as flasks and beakers. Questions included those requiring translation from 3-D drawings to conventional diagrams, as well as more simple examples testing recognition of symbols.

Mark scheme

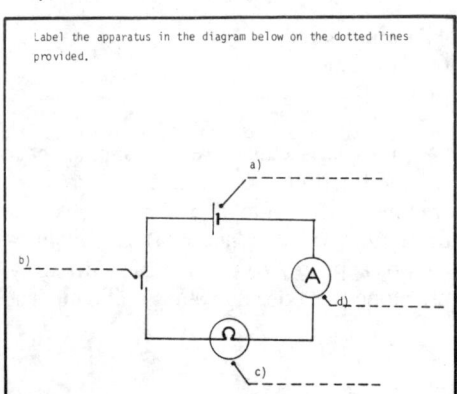

Label the apparatus in the diagram below on the dotted lines provided.

(Give credit in spite of spelling mistakes if the intention is clear)

| | | |
|---|---|---|
| a) | Cell (allow battery) | 1 |
| b) | Switch | 1 |
| c) | Lamp, or bulb | 1 |
| d) | Ammeter | 1 |

Total ④

*The mean score was 51%.*

*34% of the pupils failed to score, of whom 26% made no attempt. Only 29% of the pupils named all four components.*

'Water bath': *drawing section diagrams*

**Mark Scheme**

Look carefully at the drawing of the test tube propped up in a beaker of water. The test tube is about a quarter full of milk.

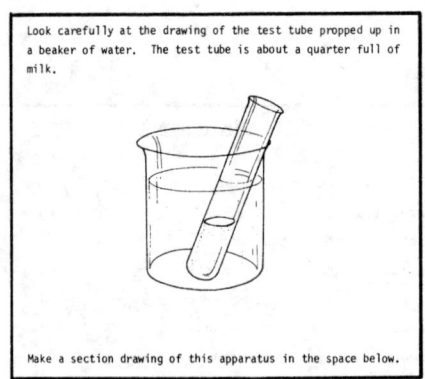

Make a section drawing of this apparatus in the space below.

(Mark only if the base of the beaker is drawn as a straight line.
Give no marks if the apparatus is not assembled)

| | |
|---|---|
| Liquid level horizontal in beaker (if the line goes across tube. (1)) | 2 |
| Liquid level horizontal in tube | 1 |
| Tube resting on beaker at top | 1 |
| at bottom | 1 |
| Reasonable proportion of height to width in t.t. | 1 |
| in beaker | 1 |
| between t.t. & beaker | 1 |
| Use of ruler where appropriate | 1 |
| Neat enough to be clear | 1 |
| Open top to test tube | 1 |
| Open top to beaker | 1 |

Total ⑫

*59% of the pupils failed to score, of whom 11% made no attempt. Although only 7% gained full marks, 34% scored 8 or more.*

Pupils appeared to have great difficulty with scientific symbols and conventions, and it may be that their use is not specifically taught by the age of 13. They are, on the other hand, commonly used in text-books, work sheets and on the blackboard.

## Summary: USE OF GRAPHICAL AND SYMBOLIC REPRESENTATION

Many kinds of charts and graphs have been used as the basis of questions, and most mean scores are between 50% and 80%. The presence of numerical data — particularly of decimal numbers — brings performance down; pupils are also less successful if two or more sets of relationships are given together, as in a question which gives both 'thinking distance' and 'braking distance' for cars moving at different speeds.

During early discussion of the assessment framework, teachers suggested that the kinds of skill represented in this category might prove a major hurdle for many pupils, preventing success in other more complex activities. It seems that things are not as bad as they feared; most pupils can read off and express information adequately in a variety of ways, unless the conventions which they need to use are specific to science. While for the simple indentification of components of circuit or section diagrams, mean scores can vary from 20% to 70%, when it comes to translating three-dimensional drawings into conventional diagrams, they are restricted to the range 20% to 30%.

# AN OVERALL VIEW OF PERFORMANCE IN SCIENCE

## The profile of performance

All the question results given in the last chapter refer to a sample of the whole population of 13 year olds, rather than an individual pupil, and are expressed as *mean scores*. It is clear what this mean score represents because the question, its mark scheme and the score distribution are on view. It is much more difficult to comprehend what general general level of performance is represented by the mean score for a whole sub-category. To say, for example, that the mean score for **Using graphs, tables and charts** is 57% makes little sense, at present, unless accompanied by a description of the sub-category and examples of questions used to assess it. As the range of questions exposed becomes greater, such statements will become more meaningful.

The sub-category results in the figure on page 40 are based on *samples* of both pupils and questions and are therefore estimates. An indication of the accuracy of the estimates is given by the **confidence intervals** which are drawn against the bars. There is a 95% probability that the actual performance level would be within the range shown. The intervals are different lengths for different sub-categories because of differences in the variety and number of questions involved as well as in the numbers of pupils and schools.

A few sub-categories of the framework do not figure in the profile. This is because too few questions could be included in the surveys to justify generalising the results.

**Estimates of % mean sub category performance of all 13-year olds in England, Wales and Northern Ireland in 1981**

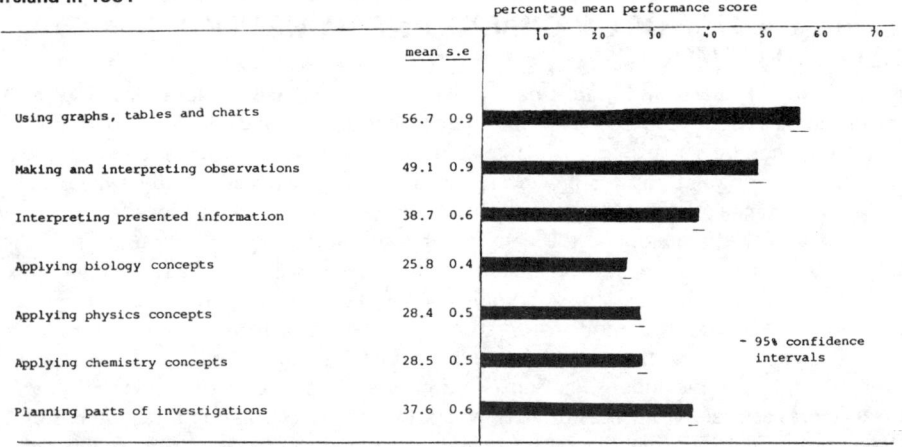

There is a large difference in mean score between **Using graphs, tables and charts** and **Applying biology concepts.** At first sight it seems as though this difference could be arbitrarily changed by altering the composition of the question bank, but this is not altogether the case. The range of difficulty within a sub-category is controlled, to a great extent, by the question-types permitted and the concept statements and contexts thought suitable for the age group. Within these limits, questions are written to cover as wide a range of difficulty as possible with a spread across that range. As a result, the system is stable: the *profile* of performance is very much the same from one year to the next, and for all three ages surveyed (even though the question banks differ across ages). It therefore seems to be a characteristic of the assessment framework rather than of a specific set of questions and mark schemes. Given the framework, the mean scores are not arbitrary values, and so differences between them have some educational significance.

Does this mean that the standard of performance in **Applying biology concepts** is lower than it should be? This is a different question, and one that is not easily answered. Teachers will make their own judgement about this after considering the question examples and the assessment framework which they represent.

**Sex differences**

In general, performance levels for boys and girls were very similar. However, there were particular groups of questions where differences did exist. For example, in some (but not all) questions which required the application of physics concepts boys did better than girls, whereas for many questions requiring observation skills the position was reversed.

The issue is a complex one, and more details can be found in the Full Reports[3,4]. Meanwhile further work is being carried out by the monitoring team and will be reported separately.

# Implications for Science Teaching

The survey results indicate some areas in which there appears to be a mis-match between pupil performance and the expectations implicit in science curriculum materials. Sometimes expectation is higher than actual performance demonstrated in the survey, and sometimes lower. Teachers may wish to consider whether the performance of the individual pupils in their care might be improved by focusing attention on these aspects as they arise in the normal course of teaching. This would allow difficulties to become apparent at an early stage so that they could be dealt with promptly and systematically.

Some teachers have expressed surprise at the relative competence of 13 year olds in using **graphs, charts and tables.** The most noticeable failure point in this area is in scaling and labelling axes of graphs. Many pupils do not use a linear scale, and it is uncommon to find axes labelled with both quantity and unit. It might improve performance to provide experience in these activities in isolation. Pupils are much less able to handle **scientific conventions,** like circuit diagrams or section diagrams of general laboratory apparatus. This is a serious disadvantage for pupils if they are to use textbooks — or more particularly worksheets — which make use of these conventions and expect their use in answers to questions. It would be worth pupils spending additional time to achieve mastery in order to smooth the way ahead.

Some **measuring instruments** are not read satisfactorily; in particular the ammeter and manometer cause difficulties, with three-quarters of the pupils failing to come within two scale divisions of the actual reading. If the teaching of science is organised on the assumption that practical experience will make it easier for pupils to develop the relevant network of scientific concepts, then it is important to put this matter right. If pupils cannot read an ammeter they will not arrive at the laws of electrolysis — or even at the fact that the current is the same all the way round a circuit — as a result of doing practical work.

**Estimating quantities** proves very difficult; there is some indication that pupils are confused over the use of the words 'mass' and 'volume'. It is possible that simple activities based on estimating — a competitive exercise, perhaps — would help to remove some of the confusion.

In **following instructions** pupils do well up to the point where they have to apply some knowledge of science. This is good news for teachers interested in the use of worksheets; very few pupils ask for help in reading the instructions, even though invited to do so; and most get on with the task quickly and competently. This is in contrast with much common experience, in which pupils reject worksheets in favour of getting their teacher to tell them what to do. Of course, the conditions in an APU test, with all necessary equipment available and in working order, and a general atmosphere of concentration, are not so easy to reproduce in the day-to-day routine of science lessons. But pupils *can* use worksheets if they see the need for it. Perhaps the messsage is 'only when necessary, only after trial and amendments, and only if suitably backed by the necessary equipment'.

Unless directed otherwise, pupils restrict their **observations** to those that can be made using sight alone. They do not, as a matter of routine, take notice of phenomena detectable by any of the other senses. Of course, it would be inadvisable to instruct pupils to taste indiscriminately but they can be taught to smell in a safe way, and

41

certainly can come to no harm in listening, or in general, in picking things up to examine them. When pupils are taught to observe systematically they begin by noting a large number of things which their teachers know are irrelevant to the purpose in hand, and this can be time-consuming; but as they gain in experience, pupils learn to discard irrelevancies; this puts them in a sounder position than that in which they disregard all senses but sight.

Pupils find it difficult to **describe a relationship** which they may perceive in information presented to them. This is true even for everyday relationships. And yet pupils are expected to generalise — to come to a conclusion — from scientific data such as results from a class experiment. What is more, they are expected to memorise the conclusion and apply it when the need arises. For example, if pupils hang various weights in turn on the end of a spring they are expected to be able to say, at the very least, "the bigger the weight the longer the spring" and in some cases "the extension is proportional to the load". Such generalisations are always 'one-off' exercises; the opportunity for pupils to try their hand at describing a relationship normally occurs only once in a lesson. And if pupils have difficulty in stating the relationship, it is usual to put the words in their mouths in the interests of progress. Perhaps it is not surprising to find that the subsequent **application** of the relationship — of Hooke's law, for example — is difficult. It may well be that "describing a relationship" is a process skill which cannot be much improved by teaching. However, inspection of curriculum and examination materials in general use suggests that, at present, little systematic attempt to encourage it is made during school science courses. Perhaps, if offered practice in the perception and use of simple relationships (not necessarily science orientated), pupils would develop the skill so that end-of-experiment 'conclusions' would become actively understood rather than passively accepted.

In performing investigations 13 year olds show a great deal of creativity and enthusiasm. Most of them are able to identify correctly both the independent and the dependent variable within an investigation, although if either is complex it is subsequently much more difficult for the pupils. In general, though, there is a lack of appreciation of the value of, or perhaps even the necessity for, **quantitative measurements.** In particular problems, such as those to do with live animals, as many as half the pupils feel able to judge for themselves without measuring anything. It might be of value to indicate, across a wide range of examples set in everyday as well as in science contexts, when the 'scientific' approach is particularly appropriate in order to arrive at reliable and accurate solutions to problems. Once the need for measurement is more clearly understood then the need for **careful measurement techniques,** which also proved to be an obstacle to success, would naturally follow.

## Conclusion

Many of the process skills in which, for the majority of 13 year olds, performance in the surveys was low appear to be basic needs of science courses of all kinds. Yet is is not common for teachers to focus on them separately and systematically, but rather to assume that pupils will pick them up in passing. This is a little like hoping children will learn multiplication tables as a result of haphazard use every other week, or will acquire a working vocabulary in French from an annual trip to Boulogne. Of course, an emphasis on drill with no further purpose is a disastrous policy, not to be recommended. But if we accept, in the interests of consolidation, the need for pupils to practise substituting figures in formulae or labelling the parts of the digestive tract, it seems reasonable to try the effect of similar systematic practice in the science process skills.

# APU science publications

Annual surveys of performance in science at ages 11, 13 and 15 will be carried out until 1984. Thereafter the frequency will drop in accordance with APU policy.

A full report is written for each survey at each age. Those for 1980 were published by HMSO; those for subsequent years will be distributed free by the APU to chief education officers, LEA science advisers, teacher training establishments, teachers' centres, examination bodies, HMI and certain other educational organisations in this country and overseas.

A separate series of shorter publications for teachers (of which this report is an example) is also being produced and each publication will focus on a different aspect of the science surveys. A copy of each of these reports for teachers will be sent free to all schools and to those receiving copies of the full survey reports. After this wide initial free distribution, copies of the science reports for teachers will be available for purchase from the Association for Science Education (see back cover for details).

*References*
Reference is made in the text to some of the publications described above, as follows:—

*Ref.*

*Science Reports for teachers*
1   The Assessment Framework at 13/15
2   Practical Assessment of Science

*Full Reports*
3   Science in Schools. Age 13: Report No. 1 DES (1980) HMSO.
4   Science in Schools. Age 13: Report No. 2

# APU Steering Group on Science

| | |
|---|---|
| Mr. A. Clegg HMI (Chairman) | APU |
| Professor P.J. Black | Director, Science Monitoring Team, Chelsea College |
| Mr. W.F. Archenhold | Director, Science Monitoring Team, Leeds University |
| Mr. N.B. Evans HMI | HM Inspectorate (Wales) |
| Mr. E.O. James | Deputy Head, Southlands School |
| Mr. J. Jeffery | Pocklington School, Pocklington |
| Professor R. Kempa | Department of Education, University of Keele |
| Dr. W.J. Kirkham | Science Adviser, Leicestershire |
| Mr. E.R.B. Little HMI | HM Inspectorate |
| Mr. I.W. Milligan | Department of Education for Northern Ireland |
| Dr. B. Prestt | Manchester Polytechnic |
| Mr. H. Wilcock | Headteacher, Paganel Junior School, Birmingham |